TRIP GENERATION

6th Edition • Volume 3 of 3

User's Guide

Institute of Transportation Engineers

Trip Generation, 6th Edition

An Informational Report of the
Institute of Transportation Engineers

User's Guide, Volume 3 of 3

The Institute of Transportation Engineers (ITE) is an international educational and scientific association of transportation and traffic engineers and other professionals who are responsible for meeting mobility and safety needs. The Institute facilitates the application of technology and scientific principles to research, planning, functional design, implementation, operation, policy development and management for any mode of transportation by promoting professional development of members, supporting and encouraging education, stimulating research, developing public awareness, and exchanging professional information; and by maintaining of a central point of reference and action.

Founded in 1930, the Institute serves as a gateway to knowledge and advancement through meetings, seminars, and publications; and through our network of approximately 15,000 members working in some 80 countries. The Institute also has more than 70 local and regional chapters and more than 90 student chapters that provide additional opportunities for information exchange, participation and networking.

Institute of Transportation Engineers
525 School St., S.W., Suite 410
Washington, D.C. 20024-2797 USA
Telephone: +1 (202) 554-8050
Fax: +1 (202) 863-5486
ITE on the Web: http://www.ite.org

Publication No. IR-016D
Fourth Printing
1,000/AGS/1198

ISBN 0-935403-09-4
Printed in the United States of America

TABLE OF CONTENTS

Volume One
Trip Generation Rates, Plots, and Equations

Port and Terminal (Land Uses 000-099)

Industrial/Agricultural (Land Uses 100-199)

Residential (Land Uses 200-299)

Lodging (Land Uses 300-399)

Recreational (Land Uses 400-499)

VOLUME TWO

TRIP GENERATION RATES, PLOTS, AND EQUATIONS

Institutional (Land Uses 500-599)

Medical (Land Uses 600-699)

Office (Land Uses 700-799)

Retail (Land Uses 800-899)

Services (Land Uses 900-999)

VOLUME THREE, USER'S GUIDE

PREFACE

Trip Generation is an Informational Report of the Institute of Transportation Engineers (ITE). The information in this document has been obtained from the research and experiences of transportation engineering and planning professionals. ITE Informational Reports are prepared for informational purposes only and do not include ITE recommendations on the best course of action or the preferred application of the data.

Trip Generation is an educational tool for planners, transportation professionals, zoning boards, and others who are interested in estimating the number of vehicle trips generated by a proposed development. This document is based on more than 3,750 trip generation studies submitted to the Institute by public agencies, developers, consulting firms, and associations.

The Sixth Edition of *Trip Generation* includes several significant changes in format and content as compared to the Fifth Edition. To facilitate the use of this document, it has been repackaged into three volumes: volumes 1 and 2 contain land use descriptions and data plots; and volume 3, the *User's Guide*, contains general introductory, instructional, and appendix material. Users are encouraged to review and become familiar with the *User's Guide* prior to using the data contained in volumes 1 and 2.

User comments on *Trip Generation* are invited. Through user feedback, the Institute has enhanced each subsequent edition of *Trip Generation*. The Institute continually seeks additional ways to increase the value of this document. ITE requests that users provide recently collected data for the land uses presented in *Trip Generation* or any other land uses for inclusion in future editions and updates. Comment and data collection forms are included in appendix B of the *User's Guide*.

ACKNOWLEDGMENTS

This Sixth Edition of *Trip Generation* is a result of many months of concerted effort by dedicated volunteers and the ITE Headquarters staff.

The Institute is particularly appreciative of the efforts put forth by Eugene D. Arnold Jr. and Brian S. Bochner. Their dedicated service, expertise, and insight contributed immensely to the ultimate completion of this publication.

Lisa M. Fontana, the Institute's Technical Program Manager for Transportation Planning, served as project manager for the publication, assembled and analyzed all the data received, conducted the statistical analyses and validation, and coordinated the project and volunteer activities. Also acknowledged are ITE staff members Michelle V. Peña for the production of the book, Cynthia de Jesus for her administrative support, and Thomas W. Brahms and Mark R. Norman for their guidance. Other staff members acknowledged for contributions to this publication include Loraine D. Coleman, Laura Hazan, and Jane Wetz.

Special thanks to Dan Schruefer, Senior Program Analyst for HBOC in Rockville, Maryland, for providing programming support that enabled modifications to the existing customized software program.

Finally, the Institute expresses its appreciation to those many agencies, firms, and individuals who have and continue to provide data to this effort.

General guidance and review of this publication were provided by members of the ITE Trip Generation Advisory Committee and Trip Generation Focus Group:

Trip Generation Advisory Committee Chairperson:
Eugene D. Arnold Jr., P.E. (F), Virginia Transportation Research Council, Charlottesville, Virginia

Trip Generation Advisory Committee:
Ahmad D. Al-Akhras, P.E. (M), Mid-Ohio R.P.C., Columbus, Ohio
Robert T. Dunphy (A), Urban Land Institute, Washington, D.C.
Frederick E. Gorove, P.E. (F), Gorove/ Slade Associates, Inc., Washington, D.C.
Kevin G. Hooper, P.E. (M), T.Y. Lin International, Falmouth, Maine
Michele W. Johnson, Federal Highway Administration, Baltimore, Maryland
Robert P. Jurasin, P.E. (F), Wilbur Smith Associates, New Haven, Connecticut
Rolf P. Kilian (M), Metro Transportation Group, Inc., Hanover Park, Illinois
Chris D. Kinzel, P.E. (FL), TJKM Transportation Consultants, Pleasanton, California

Kenneth G. Mackiewicz, P.E. (M), Raymond Keyes Associates Inc., Tarrytown, New York

David Muntean, Jr., P.E. (M), Transportation Consulting Group, Tallahassee, Florida

Allyn D. Rifkin, P.E. (M), City of Los Angeles Department of Transportation, Los Angeles, California

George E. Schoener, US DOT-Federal Highway Administration, Washington, D.C.

Kent A. Whitson, P.E. (M), San Diego Association of Governments, San Diego, California

Trip Generation Focus Group:

Robert E. Bartlett, P.E. (M), Hannover, Germany

Daniel L. Cronin, P.E. (F), Balloffet & Associates, Inc., Denver, Colorado

G. Bruce Douglas, P.E. (M), PBQD, Herndon, Virginia

Slade F. Exley, P.E. (M), Neel-Schaffer, Inc., Jackson, Mississippi

Laura B. Firtel, AICP (A), Kimley-Horn & Associates, Inc., Orlando, Florida

Jon D. Fricker, P.E. (M), Purdue University, West Lafayette, Indiana

James W. Gough, P. Eng. (M), Marshall Macklin Monaghan, Ltd., Thornhill, Ontario, Canada

Richard C. Hawthorne, P.E. (M), Maryland National Capital P.P.C., Silver Spring, Maryland

Shelly A. Johnston, P.E. (A), Transportation Concepts, Clifton Park, New York

James H. Kawamura, P.E., KHR Associates, Irvine, California

Wayne K. Kittelson, P.E. (M), Kittelson & Associates, Inc., Portland, Oregon

Jim C. Lee, P.E. (F), Lee Engineering, Inc., Phoenix, Arizona

Robert E. Leigh, P.E. (ML), Leigh, Scott & Cleary, Inc., Denver, Colorado

Richard W. Lyles, P.E. (F), Michigan State University, East Lansing, Michigan

Robert W. McBride, P Eng (M), BA Consulting Group Ltd., Toronto, Ontario, Canada

Alan Paling, JMP Consultants, Ltd., Lincoln, England

Shiva K. Pant (M), Fairfax County Department of Transportation, Fairfax, Virginia

William J. Roache, P.E. (M), Vanasse Hangen Brustlin, Inc., Watertown, Massachusetts

Jack S. Schnettler, P.E. (M), Post-Buckley-Schuh-Jernigan, Miami, Florida

Sheldon Schumacher, P.E. (FL), Homart Development Company, Chicago, Illinois

John L. Simon (M), The Taubman Company, Inc., Bloomfield Hills, Michigan

Gary H. Sokolow (M), Florida Department of Transportation, Tallahassee, Florida

Robert E. Stammer, P.E. (F), Vanderbilt University, Nashville, Tennessee

Steven A. Tindale, P.E. (M), Tindale-Oliver & Associates, Inc., Tampa, Florida

Alan L. Tinter, P.E. (F), Tinter Associates, Inc., Fort Lauderdale, Florida

Raymond S. Trout, P.E. (F), A/E Group, Inc., Westminster, Maryland

Michael W. Van Aerde, P Eng (M), Queen's University, Kingston, Ontario, Canada

Richard P. Wolsfeld, P.E. (F), BRW, Inc., Minneapolis, Minnesota

Robert C. Wunderlich, P.E. (M), City of Garland Transportation Department, Garland, Texas

David A. Younger, P.E. (M), City of Columbus, Columbus, Ohio

Bassam A. Ziada (A), Walmart Stores, Inc., Bentonville, Arkansas

(Letters in parentheses indicate ITE member grade: A-Associate Member, M-Member, F-Fellow, FL-Fellow Life, ML-Member Life.)

INTRODUCTION

Purpose

The purpose of this multivolume document is to house the trip generation data that have been submitted to ITE. This document represents the sixth full edition, and it incorporates data from the previous five editions and the supplementary document entitled *February 1995 Update to the Fifth Edition*. As additional trip generation data become available, they will be distributed through the periodic publication of updates.

Use of the Report

This document is intended for use in estimating the number of trips that may be generated by a specific land use. Trip generation rates and equations have been developed for an average weekday, Saturday, and Sunday; for the weekday morning and evening peak hours of the generator; and for the weekday morning and evening peak hours of the adjacent street traffic, i.e., between 7 A.M. and 9 A.M. and 4 P.M. and 6 P.M. In some cases limited data were available; thus, the statistics presented may

not be truly representative of the trip generation characteristics of a particular land use. Further information on the cautions and limitations of the data contained in this report are presented in chapter 4, "Description of the Data Base."

About the Data

The average trip generation rates in this document represent weighted averages of studies collected throughout the United States and Canada since the 1960s. Data were primarily collected at suburban locations with little or no transit service, near-by pedestrian amenities, or transportation demand management (TDM) programs. At specific sites, the user may wish to modify trip generation rates presented in this document to reflect the presence of public transportation service, ridesharing or other TDM measures, enhanced pedestrian and bicycle trip-making opportunities, or other special characteristics of the site or surrounding area. When practical, the user is encouraged to supplement the data in this document with local data that have been collected at similar sites. Additional collected data should be submitted to the Institute for inclusion in subsequent editions. Data collection forms are provided in appendix B of the *User's Guide*. Questions and comments regarding *Trip Generation* should be addressed to:

Institute of Transportation Engineers
Programs & Services Department
525 School Street, S.W., Suite 410
Washington, D.C. USA 20024-2797
Telephone: +1 202/554-8050
Fax: +1 202/863-5486
ITE on the Web: http://www.ite.org

▲2 CHANGES SINCE THE FIFTH EDITION

The Sixth Edition of *Trip Generation* includes several significant changes in format and content as compared to the Fifth Edition. To facilitate the use of this document, it has been repackaged into three volumes: volumes 1 and 2 contain land use descriptions and data plots; and volume 3, the *User's Guide*, contains general introductory, instructional, and appendix material. Nineteen new land use classifications and data from over 750 sites have been added in the Sixth Edition. In addition, revisions have been made to the instructional information and to several land use codes, independent variables, and land use definitions. Specific changes are described in the following sections.

Format

- Because of the magnitude of the information collected and the resulting size of the publication, the Sixth Edition has been published as three volumes. Volume 1 contains trip generation rates, plots, and equations for the Port and Terminal, Industrial/ Agricultural, Residential, Lodging, and Recreational land uses (land uses 000-499). Volume 2 includes the same information for the Institutional, Medical, Office, Retail, and Services land uses (land uses 500-999). For user convenience, the introductory and instructional material as well as the appendices are included in the third volume, the *User's Guide*.

- The table of contents has been expanded to include the numerical listing of land uses previously contained in appendix A of the Fifth Edition.

- The tables presenting land uses with only one data point have been moved to the beginning of each land use.

Land Use Codes

A significant amount of new data has been collected since the publication of the Fifth Edition. Data from more than 750 new sites have been added to the existing data base for a combined total of more than 3,750 individual trip generation studies. Data collection efforts have resulted in the addition of nineteen new land uses:

- Light Rail Transit Station with Parking (093)

- Miniature Golf Course (431)

- Golf Driving Range (432)

- Multipurpose Recreational Facility (435)

- Automobile Racetrack (453)

- Ice Rink (465)

- Casino/Video Lottery Establishment (473)

- Middle School/Junior High School (522)

- Free-Standing Discount Superstore (813)

- Fast-Food Restaurant with Drive-Through Window and No Indoor Seating (835)

- Quick Lubrication Vehicle Shop (837)

- Automobile Parts Sales (843)

- Wholesale Tire Store (849)

- Home Improvement Superstore (862)

- Electronics Superstore (863)

- Toy/Children's Superstore (864)

- Pharmacy/Drugstore without Drive-Through Window (880)

- Pharmacy/Drugstore with Drive-Through Window (881)

- Video Rental Store (896)

Further, based on new information, some land uses were renamed, split into multiple uses, or deleted. Following is a list of the changes that were made:

- Bus Park and Ride Station (090) was renamed Park-and-Ride Lot with Bus Service (090) to more accurately describe the studies contained therein;

- Retail General Merchandise (810) was deleted because the existing land use

definition was too broad for site-specific application;

- Discount Store (815) was changed to Free-Standing Discount Store (815) to more accurately describe the studies contained therein;

- Drinking Place (835) was changed to land use code 836;

- Service Station (844); Service Station with Convenience Market (845), and Service Station with Convenience Market and Car Wash (846) have been changed to include "gasoline" in the land use name in order to more accurately describe the studies therein;

- Walk-In Savings and Loan (913) and Drive-In Savings and Loan (914) were deleted as they are no longer viable land uses.

Land Use Descriptions

- Building occupancy data were added to the land use description pages for the following land uses: Hotel (310), All Suites Hotel (311), Motel (320), Resort Hotel (330), and General Office Building (710).

- A caution concerning the trip generation statistics was added to several of the park land uses. It was noted that the percentage of the park acreage that is actually used varies considerably among the sites in the data base. This caution applies to City Park (411), County Park (412), State Park (413), and Regional Park (417).

- A caution was added concerning the Saturday and Sunday trip generation statistics at a Junior/Community College (540) and at a University/College (550). It was noted that there was a significant variation among the sites in the data base.

- A caution was added concerning the use of weekday statistics to represent a typical Friday generation at a Movie Theater without Matinee (443) and a Movie Theater with Matinee (444).

- The definition for Building Materials and Lumber Store (812) was refined to include only buildings that are less than 25,000 gross square feet in size. The new land use Home Improvement Superstore (862) includes similar stores that are larger in size.

- The definition of Free-Standing Discount Store (815) was revised to omit the reference to mutual operation with supermarkets. Free-standing discount stores containing full service grocery departments are now contained in Free-Standing Discount Superstore (813).

- The definition of Shopping Center (820) was expanded to note that the studies contained within this land use include neighborhood, community, regional, and super regional centers.

- The definitions for Quality Restaurant (831) and High Turnover (Sit-Down) Restaurant (832) were clarified by noting that quality restaurants usually require reservations whereas high turnover sit-down restaurants do not. Also, it was emphasized that quality restaurants are not typically chains.

- The definitions for High Turnover (Sit-Down) Restaurant (832), Fast-Food Restaurant without Drive-Through Window (833), and Fast-Food Restaurant with Drive-Through Window (834) were changed to note that the sites contained in the data base may or may not have been open for breakfast; thus, users should exercise caution when applying the trip generation statistics during the A.M. period.

- The definition for High Turnover (Sit-Down) Restaurant (832) was modified to include turnover rates of approximately one hour or greater to more accurately describe the studies contained in the data base. The definition was also changed to indicate that some facilities in this land use may contain a bar area for serving food and alcoholic drinks.

- The names and definitions for the service station land uses (844, 845, 846) were changed to clarify that they include facilities for fueling motor vehicles, but they may or may not include facilities for servicing and/or repairing motor vehicles.

- The definitions of Service Station with Convenience Market (845), Service Station with Convenience Market and Car Wash (846), and Convenience Market with Gasoline Pumps (853) were clarified. Land uses 845 and 846 describe service stations where the primary business is the fueling of motor vehicles. Land use 853 describes convenience markets with gasoline pumps where the primary business is the sale of convenience items.

- The definition of Tire Store (848) was clarified to indicate that these stores do not generally contain large storage or warehouse areas. The new land use Wholesale Tire Store (849) includes tire stores that contain significant storage or warehouse areas.

- The definition of Supermarket (850) was modified to indicate that super-markets may include facilities such as money machines, photo centers, pharmacies, and video rental areas.

Independent Variables

The following list summarizes additions or changes made to the independent variables. See chapter 3, "Definition of Terms," for more detailed descriptions of the variables identified in this section.

- An explanation of the relationship between the independent variables "gross floor area" (GFA), "gross leasable area" (GLA), and "gross rentable area" (GRA) was added.

- The A.M. and P.M. peak hours of the adjacent street traffic volume were added as independent variables to Fast-Food Restaurant with Drive-Through Window (834) and to some of the service station and gasoline-related land uses (844, 845, 853).

- The independent variable for all service stations and gasoline-related land uses (844, 845, 846, 853) was changed from the number of pumps and/or hoses to

the number of vehicle fueling positions (VFP).

Definition of Terms

- The definition of trip generation during the **peak hour of adjacent street traffic** was refined to help the user better understand and apply the statistics reported in *Trip Generation*. See chapter 3, "Definition of Terms," for further information.

Other

- The regression analysis procedures in this edition were modified as follows (see chapter 5, "Description of Data Plots and Reported Statistics"): (1) The regression equations include only the linear and logarithmic equations. The inverse, linear-logarithmic, and logarithmic-linear equations were removed in order to more simply and logically display the characteristics of the given data. (2) The regression equation with the highest **coefficient of determination, R^2,** was always plotted. In previous editions, the linear model was selected if its R^2 was no more than 0.05 less than that of the model with the highest R^2. (3) Best fit regression equations and curves were shown only when the R^2 value was greater than or equal to 0.50. The threshold for R^2 was 0.25 in the Fifth Edition and 0.40 in the *Fifth Edition Update*.

- For some land uses there were fewer studies reported in the Sixth Edition than in the Fifth Edition. Several factors accounted for this: certain studies were shifted because of the creation of new land use classifications; duplicate studies were deleted; or studies were deleted as a result of inconsistencies in the reported data. In particular, several studies were omitted from Fast-Food Restaurant without Drive-Through Window (833) and Fast-Food Restaurant with Drive-Through Window (834). The only sites now included in these land uses are those that clearly identify whether or not a drive-through facility was present.

- For some land uses the directional distributions changed despite the fact that the sites used in these land uses have remained the same. This variation is a result of the reexamination of information received for these studies and the resulting recalculation of the directional distributions.

- The land use density tables previously provided on the land use description pages were deleted from the Sixth Edition. The information was deemed inappropriate as it was not representative of the entire set of data reported for the specific land use.

- Plots containing blow-ups of subsets of data were removed because they provide no useful additional information.

- The tables that reported adjustment factors for average weekday vehicle trip rates for residential land uses were removed due to the age of the data source. Subsequent studies to update this information have not been conducted to date.

- Average rates and equations were added to the plots for General Office Building (710) and Shopping Center (820) to maintain consistency with the reporting of information for the other land uses.

- The data for Shopping Center (820) were no longer separated into subsets based on the size of the shopping center in order to maintain consistency with the reporting of data for the other land uses.

- Forms for submitting data to ITE were revised to request additional information and to more clearly define the statistical data. Copies of these forms are included in appendix B of the *User's Guide.*

DEFINITION OF TERMS

The **average trip rate** is the weighted average of the number of vehicle trips or trip ends per unit of independent variable (e.g., trip ends per occupied dwelling unit or employee) using a site's driveway(s). The weighted average rate is calculated by summing all trips or trip ends and all independent variable units where paired data are available, and then dividing the sum of the trip ends by the sum of the independent variable units. The weighted average rate is used rather than the average of the individual rates because of the variance within each data set or generating unit. Data sets with a large variance will over-influence the average rate if they are not weighted.

The **average trip rate for the peak hour of the adjacent street traffic** is the one-hour weighted average vehicle trip generation rate at the site between 7 A.M. and 9 A.M. and between 4 P.M. and 6 P.M., when the combination of its traffic and the traffic on the adjacent street is the highest. If the adjacent street traffic volumes are unknown, the average trip rate for the peak hour of the adjacent street represents the highest hourly vehicle trip ends generated by the site during the traditional commuting peak periods of 7 A.M. to 9 A.M. and 4 P.M. to 6 P.M.

The **A.M. and P.M. peak hour volume of adjacent street traffic** is the highest hourly volume of traffic on the adjacent street during the A.M. and P.M., respectively.

The **average trip rate for the peak hour of the generator** is the weighted average vehicle trip generation rate during the hour of highest volume of traffic entering and exiting the site during the A.M. or the P.M. hours. It may or may not coincide in time or volume with the trip rate for the peak hour of the adjacent street traffic. The trip rate for the peak hour of the generator will be equal to or greater than the trip rate for the peak hour between 7 A.M. and 9 A.M. or between 4 P.M. and 6 P.M.

The **average weekday vehicle trip ends (AWDVTE)** is the average 24-hour total of all vehicle trips counted to and from a study site from Monday through Friday.

The **average weekday trip rate** is the weighted weekday (Monday through

Friday) average vehicle trip generation rate during a 24-hour period.

The **average Saturday trip rate** is the weighted average Saturday vehicle trip generation rate during a 24-hour period.

The **average trip rate for the Saturday peak hour of the generator** is the weighted average Saturday vehicle trip generation rate during the hour of highest volume of traffic entering and exiting a site. It may occur in the A.M. or P.M.

The **average Sunday trip rate** is the weighted average Sunday vehicle trip generation rate during a 24-hour period.

The **average trip rate for the Sunday peak hour of the generator** is the weighted average Sunday vehicle trip generation rate during the hour of highest volume of traffic entering and exiting a site. It may occur in the A.M. or P.M.

The **gross floor area (GFA)**[1] of a building is the sum (in square feet) of the area of each floor level, including cellars, basements, mezzanines, penthouses, corridors, lobbies, stores and offices, that are within the principal outside faces of exterior walls, not including architectural setbacks or projections. Included are all areas that have floor surfaces with clear standing head room (6 feet, 6 inches minimum) regardless of their use. If a ground-level area, or part thereof, within the principal outside faces of the exterior walls is not enclosed, this GFA is considered part of the overall square footage of the building. However, unroofed areas and unenclosed roofed-over spaces, except those contained within the principal outside faces of exterior walls, should be excluded from the area calculations. For purposes of the trip generation calculation, the GFA of any parking garages within the building should not be included within the GFA of the entire building. The majority of the land uses in this document express trip generation in terms of GFA. In *Trip Generation*, the unit of measurement for office buildings is currently GFA; however, it may be desirable to also obtain data related to gross rentable area and net rentable area. With the exception of buildings containing enclosed malls or atriums, gross floor area is equal to gross leasable area and gross rentable area.

The **gross leasable area (GLA)**[2] is the total floor area designed for tenant occupancy and exclusive use, including any basements, mezzanines, or upper floors, expressed in square feet and measured from the centerline of joint partitions and from outside wall faces. For purposes of the trip generation calculation, the floor area of any parking garages within the building should not be included within the GLA of the entire building. GLA is the area for which tenants pay rent; it is the area that produces income. In the retail business, GLA lends itself readily to measurement

---

[1] Institute of Real Estate Management of the National Association of Realtors. *Income/Expert Analysis, Office Buildings, Downtown and Suburban*, 1985, p. 236.
[2] Urban Land Institute. *Dollars and Cents of Shopping Centers*, 1984.

and comparison; thus, it has been adopted by the shopping center industry as its standard for statistical comparison. Accordingly, GLA is used in this report for shopping centers. For strip centers, discount stores and freestanding retail facilities, GLA usually equals GFA.

The **gross rentable area (GRA)**[3] is computed in square feet by measuring the inside finish of permanent outer building walls or from the glass line where at least 50 percent of the outer building wall is glass. GRA includes all the area within outside building walls excluding stairs, elevator shafts, flues, pipe shafts, vertical ducts, balconies, and air conditioning rooms.

The **net rentable area (NRA)**[4] is computed in square feet by measuring inside the finish of permanent outer building walls or from the glass line where at least 50 percent of the outer building wall is glass. NRA includes all the area within outside building walls excluding stairs, elevator shafts, flues, pipe shafts, vertical ducts, balconies, air-conditioning rooms, janitorial closets, electrical closets, washrooms, public corridors, and other such rooms not actually available to tenants for their furnishings or to personnel and their enclosing walls. No deductions should be made for columns and projections necessary to the building. Typically, the NRA for office buildings is approximately equal to 85 to 90 percent of the GFA.

An **independent variable** is a physical, measurable, or predictable unit describing the study site or generator that can be used to predict the value of the dependent variable (trip ends). Some examples of independent variables used in this book are GFA, employees, seats, and dwelling units.

A **servicing position** is defined by the number of vehicles that can be serviced simultaneously at a quick lubrication vehicle shop or other vehicle repair shop. That is, if a quick lubrication vehicle shop has one service bay that can service two vehicles at the same time, the number of serving positions would be two.

A **trip** or **trip end** is a single or one-direction vehicle movement with either the origin or the destination (exiting or entering) inside a study site. For trip generation purposes, the total trip ends for a land use over a given period of time are the total of all trips entering and all trips exiting a site during a designated time period.

A **vehicle fueling position** (VFP) is defined by the number of vehicles that can be fueled simultaneously at a service station. For example, if a service station has two product dispensers with three hoses and grades of gasoline on each side, where only one vehicle can be serviced at a time on each side, the number of vehicle fueling positions would be four.

---

[3] Institute of Real Estate Management of the National Association of Realtors. *Income/Expert Analysis, Office Buildings, Downtown and Suburban*, 1985, p. 236.
[4] Ibid.

▲ 4

DESCRIPTION OF THE DATA BASE

The data analyzed in this document were contributed on a voluntary basis by various state and local governmental agencies, consulting firms, individual transportation professionals, universities and colleges, developers, associations, and local sections of the Institute of Transportation Engineers. In many cases, the data were originally contained in published reports or unpublished analyses conducted by such groups. The sources of these reports or analyses have been listed in appendix A. The source numbers for studies contained in each land use have been listed on the land use description pages.

No original field surveys were conducted by ITE Headquarters. The amount of data submitted for an individual site varied from as little as one peak hour volume to seven days of directional hourly volumes. All data have been combined to maximize the size of the data base for each land use and each time period. Data received were first examined by ITE staff for validity and reasonableness before being entered into the comprehensive data base.

Data Collection

Some of the data submitted were collected using automatic counters configured to count vehicular traffic entering and exiting a site. Presumably, the sites selected for counting did not include through traffic, and counts were taken on driveways of sufficient length to avoid the double-counting of turning vehicles. In some cases, counts were nondirectional and therefore did not distinguish between entering and exiting vehicles. Manual counts often supplemented the automatic counts to obtain vehicle occupancy and classification, check the reliability of the automatic counters, and obtain directional counts during peak periods when a nondirectional automatic count was being conducted. In other cases, only manual counts were conducted during peak periods. *All data presented in this report represent **VEHICLE** trip generation rather than person trip generation.*

Additional information regarding site characteristics was obtained through personal interviews, actual measurements, telephone conversations, or mail-back questionnaires.

Data Analysis and Storage

The statistical analyses conducted for the Sixth Edition of *Trip Generation* were derived from a statistical software program and from a trip generation data base located at ITE Headquarters. Each data record was referenced in the data base by a source number, the month and year of the traffic volume measurement, the metropolitan area (when known), and a three-digit land use code. There are now data for 136 land uses classified within ten major land use categories. Additional land uses are continuously added to the data base as data become available.

Data Age

The data base compiled to produce this document contains data extending back to the early 1960s. Consequently, there was concern that the data collected before the first major energy crisis in 1973 may differ from the post-energy crisis data. The Federal Highway Administration (FHWA) analyzed the data base from the Third Edition (1982) of *Trip Generation* and reported that "Based on statistical tests such as T-tests and F-ratios, it was concluded that there were no significant differences between the mean trip rates of older data (pre-1973) and new data (post-1973) for all land uses analyzed."[1] ITE staff performed additional analyses comparing pre- and post-1980 data for the restaurant land uses (831, 832, 833) for the *February 1995 Update to the Fifth Edition*.

Again it was found that there were no significant differences between the mean trip rates of the older data and the newer data. Thus, all data points were retained in the data base to maximize the sample sizes of the given land uses. It is anticipated that additional analyses will be performed for future updates to continue monitoring variations based on the age of the data.

Variations in the Statistics

Variations in trip generation characteristics for specific land uses are reflected in the range of rates, standard deviation, and coefficient of determination (R^2) value. (See chapter 5, "Description of Data Plots and Reported Statistics" for additional details on these topics.) These variations may be due to a small sample size, the individual marketing of the site, economic conditions of the business market, the geographic location of the sites studied, or the unique characteristics of the specific site. Accordingly, judgment must be exercised in the use of the statistics in this report.

Other sources of variation include different lengths of traffic count duration and the time of the year the traffic volumes were counted; that is, daily and seasonal variations may exist for some land uses. Studies have not been undertaken to analyze differences based on geographic location.

---

[1] Kellerco, *Development and Application of Trip Generation Rates.* U.S. Department of Transportation, FHWA, January 1985.

Limitations of the Data Plots

The plots presented in *Trip Generation* cover only the range of independent variables for which data are available. Caution should be used if extrapolating the data beyond the ranges provided, since no information has been supplied to document trip generation characteristics beyond the given ranges.

It should also be noted that in some cases, because of the limited sample size and variation in the data received, the projected trip generation estimate for the peak hour of the adjacent street traffic exceeds the trip generation estimate for the peak hour of the generator. By definition, this is impossible. In these isolated cases, knowledge of the project site and engineering judgment should be used to select the appropriate trip generation approximation.

5

DESCRIPTION OF DATA PLOTS AND REPORTED STATISTICS

Data Plots

Figure V-1 is an example of the statistical and descriptive information available for the majority of the land uses contained in volumes 1 and 2 of the Sixth Edition of *Trip Generation*. This sample data page provides explanatory notes describing each element of the figure.

Data plots provide the most fundamental display of the variance within the data base. It should be emphasized that the data points represented on the plots are not trip generation rates; rather, they are the observed number of trips, plotted against the size of the independent variable.

Data plots were not provided for land uses containing only one study for an independent variable and time of day combination. In these cases, the data were presented in tabular form immediately following the land use description page at the beginning of each land use. A

description of the equations contained on the data plots is contained in chapter 6, "Instructions."

Reported Statistics

Average Trip Rate

The average trip generation rates displayed in this report were calculated on the basis of a weighted **average trip rate**. The weighted average trip rate was used rather than the average of the individual rates because of the variance found within each data set. Sites with a large variance from the mean would have over-influenced the average rate had they not been weighted.

Standard Deviation for the Weighted Average Trip Rate

The **standard deviation** is a measure of how widely dispersed the data points are around the calculated average. The lower the standard deviation, meaning the less

Figure V-I: Sample Data Page

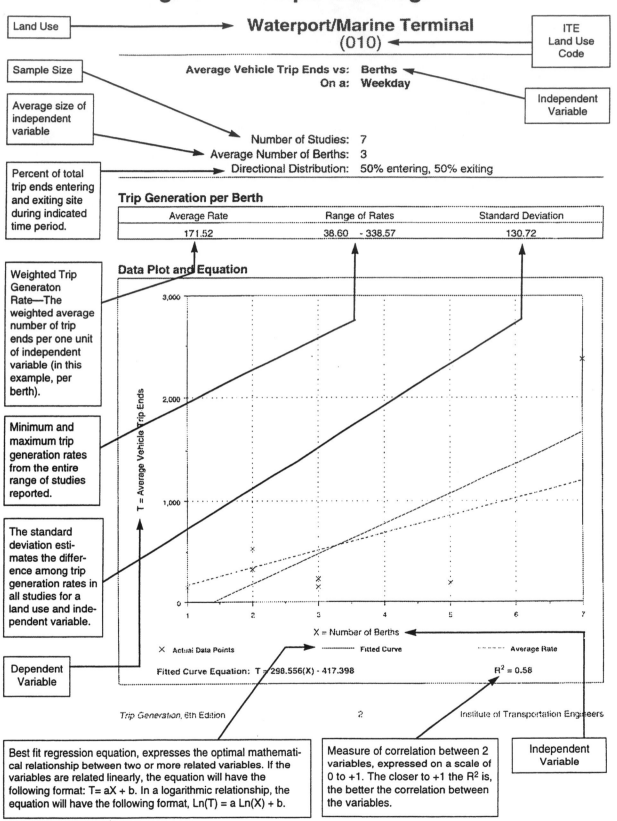

Land Use → **Waterport/Marine Terminal
(010)** ← ITE Land Use Code

Sample Size

Average size of independent variable

Percent of total trip ends entering and exiting site during indicated time period.

Average Vehicle Trip Ends vs: Berths
On a: Weekday

← Independent Variable

Number of Studies: 7
Average Number of Berths: 3
Directional Distribution: 50% entering, 50% exiting

Trip Generation per Berth

Average Rate	Range of Rates	Standard Deviation
171.52	38.60 - 338.57	130.72

Weighted Trip Generaton Rate—The weighted average number of trip ends per one unit of independent variable (in this example, per berth).

Minimum and maximum trip generation rates from the entire range of studies reported.

The standard deviation estimates the difference among trip generation rates in all studies for a land use and independent variable.

Dependent Variable

Data Plot and Equation

T = Average Vehicle Trip Ends

3,000

2,000

1,000

0

1   2   3   4   5   6   7

X = Number of Berths

✕   Actual Data Points   Fitted Curve   Average Rate

Fitted Curve Equation: T = 298.556(X) - 417.398 $R^2 = 0.58$

Best fit regression equation, expresses the optimal mathematical relationship between two or more related variables. If the variables are related linearly, the equation will have the following format: T= aX + b. In a logarithmic relationship, the equation will have the following format, Ln(T) = a Ln(X) + b.

Measure of correlation between 2 variables, expressed on a scale of 0 to +1. The closer to +1 the R^2 is, the better the correlation between the variables.

Independent Variable

Trip Generation, 6th Edition   2   Institute of Transportation Engineers

dispersion there is in the data, the better the data fit. In this document, the statistics reported are based on a "weighted average," not an "arithmetic average;" and therefore, the standard deviation is an approximation and not statistically correct.

Regression Analysis

The software used in the Sixth Edition of *Trip Generation* examines the independent variable and the number of trips in order to generate a regression curve, a regression equation, and a **coefficient of determination (R^2)** for each land use. The coefficient of determination is defined as the percent of the variance in the number of the trips associated with the variance in the size of the independent variable. If the R^2 value is 0.75, then 75 percent of the variance in the number of trips is accounted for by the variance in the size of the independent variable. As the R^2 value increases towards 1.0, the better the fit; as the R^2 value decreases towards 0, the worse the fit. A standard formula for calculating R^2 can be found in a statistics textbook.

The general forms of the regression equations used in this report include:

$$T = aX + b \text{ (linear)}$$

$$Ln(T) = aLn(X) + b \text{ (logarithmic)}$$

The objective in developing the relationship between X (the independent variable) and T (the dependent variable or number of trips) is to determine the values of the parameters "a" and "b." As a result, the expected error in estimating the dependent variable (the number of trips) given the estimates of the independent variable will be minimized.

The software program selects and plots the regression equation with the highest R^2 value. The regression equation appears on the graph as a solid line to show how well it represents the actual data points.

Best fit regression curves are shown in this document only when each of the following three conditions are met:

1. the R^2 is greater than or equal to 0.50,

2. the sample size is greater than or equal to 4,

3. the number of trips increases as the size of the independent variable increases.

INSTRUCTIONS

Trip Generation provides the user community with three methods of estimating trips at proposed developments:

1. a plot of trip ends versus size of the independent variable for each study, which can be used to graphically obtain a rough estimate of trips;

2. the weighted average trip generation rate (number of weighted trip ends per unit of the independent variable);

3. a regression equation relating trip ends to the size of the independent variable.

Understanding the Methodologies

Selecting an appropriate method for estimating trips requires the use of engineering judgment and a thorough understanding of the three methodologies listed above. The methodologies are explained in the following sections.

Graphic Plot

The most fundamental display of available information is a plot of the total trip ends versus a related independent variable.

This plot can be used to predict the number of trip ends generated for a given independent variable based on the existing data points. This method is reasonably accurate if there are sufficient data points within the range of the independent variable being used to define a relationship between the two variables. Otherwise, the need for interpreting the data (e.g., discarding "erratic" data points) and for interpolating between data points may result in inconsistent interpretations of the data.

Weighted Average Trip Rate

The traditional method of forecasting trips has been to apply a weighted average trip rate. For example, the number of trips can be estimated by multiplying the number of trip ends per unit of independent variable by the number of units of the independent variable associated with the proposed development.

The standard deviation provides a measure of how widely dispersed the data points are around the calculated average; the less the dispersion (meaning the lower

the number) the better the approximation. The approximated standard devia-
tions are provided for plots with three or
more data points. Graphically, use of the
weighted average rate assumes a linear
relationship passing through the origin
with a slope equal to the rate.

Regression Equation

Regression analysis provides a tool for
developing an equation that defines the
line that "fits best" through the data
points.

Use of the regression equation allows a
direct forecasting of trip ends based on
the independent variable of the proposed
development, thus eliminating differences
of opinion arising from interpolating a
plot of individual data points. Unlike the
weighted average rate, the plotted
equation does not necessarily pass
through the origin, nor does the
relationship have to be linear.

The correlation coefficient (R) is a
measure of the degree of association or
closeness between variables. The coefficient of determination (R^2) is the percent
of the variance in the number of the trips
associated with the variance in the size of
the independent variable. Thus, an R
value of 0.8 results in an R^2 of 0.64, which
is to say that 64 percent of the variance in
the number of trips is accounted for by
the variance in the size of the independent variable. The closer the R^2 value is to
1.0, the better the relationship between
the number of trips and the size of the
independent variable.

Sample Problem

The method of calculating trip generation
through the use of either a regression
equation or the weighted average trip
generation rate is illustrated by the sample
problem below.

For a Waterport/Marine Terminal (land
use 010) with three berths, the calculation
of the average vehicle trip ends per berth
on an average weekday is provided as
follows. Refer to the data and plot
presented for this land use in Figure V-I,
"Sample Data Page," shown in chapter 5.
The rate and equation are listed
accordingly:

Rate: T = 171.52 trip ends per berth

Equation: T = 298.556(X) - 417.398

Calculate vehicle trip ends using the rate:

T = 171.52 x 3 = 515 vehicle trip ends

Calculate vehicle trip ends using the
equation:

T = 298.556(3) - 417.398
= 478 vehicle trip ends

Choice of Day and Time Period

The day and time period that should be
used in determining the appropriate
design requirements for the proposed
development is directly related to the type
of land use and the traffic characteristics
on the adjacent street system. Trip
generation for different days and time
periods should be examined to determine
when the site being planned experiences
its peak traffic flow and to define the
relationship between the site's peak

generation and the peaking characteristics of the adjacent street system.

In most cases, the traffic volume generated by the site, combined with the traffic volume already on its adjacent street, is highest during the traditional commuting peak hours. Thus, the maximum impact would be evaluated by adding the generator peak traffic volume and the adjacent street peak traffic volume.

Some land uses, however, do not peak at the same time as the adjacent streets. Therefore, combinations of site volumes and street volumes at different potential peak times should be checked to determine the proposed site's maximum impact.

More detailed information than is included in this document may be required to determine the peak time and volumes needed for the analysis.

UPDATE PROCEDURE

ITE has established a procedure for updating the data summarized in this report and invites all interested parties to collect data from one or more sites and submit them to ITE Headquarters.

The data analyses in this report were performed using commercial software. ITE has a sophisticated system of data entry to facilitate updating the data base. This procedure will result in a continual, uniform method of obtaining and summarizing the current trip generation data for all land uses. The Institute will do the following:

- store all trip generation data;

- encourage ITE district and section technical committees, ITE student chapters, governmental agencies, and private consultants to collect additional data;

- distribute trip generation data forms to those collecting data;

- maintain a computer data base for trip generation analyses and summarization;

- maintain and modify when necessary a uniform procedure for collecting data;

- summarize trip generation data;

- conduct special trip generation analyses when appropriate;

- revise trip generation rates, equations, plots, and text on the basis of additional data;

- identify data collection needs in areas where deficiencies exist or where little information is available.

Standard data collection forms are included in appendix B of the *User's Guide* and should be returned to the Institute at the following address:

Institute of Transportation Engineers
Programs & Services Department
525 School Street, S.W., Suite 410
Washington, D.C. 20024-2797 USA
Phone: +1 202/554-8050
Fax: +1 202/863-5486
ITE on the Web: http://www.ite.org

APPENDIX A:
SOURCES

1. *Trip Generation*, Western Section, ITE, January 1967.

2. *Trip Generation Study of Selected Commercial and Residential Developments*, Illinois Section, ITE.

3. *Trip Generation for Commercial and Industrial Development*, Southern Section, ITE, 1972.

4. *Trip Generation Study*, New England Section ITE, July 1973.

5. *Trip Generation Study*, Ohio Section ITE, November 1973.

6. *Progress Report on Trip Ends Generation Research Counts*, San Francisco, CA: State of California Transportation Agency, Department of Public Works, Division of Highways, District 4, December 1965.

7. *Second Progress Report on Trip Ends Generation Research Counts*, San Francisco, CA: State of California Transportation Agency, Department of Public Works, Division of Highways, District 4, December 1966.

8. *Third Progress Report on Trip Ends Generation Research Counts*, San Francisco, CA: State of California Transportation Agency, Department of Public Works, Division of Highways, District 4, December 1967.

9. *Fourth Progress Report on Trip Ends Generation Research Counts*, San Francisco, CA: State of California Transportation Agency, Department of Public Works, Division of Highways, District 4, December 1968.

10. *Fifth Progress Report on Trip Ends Generation Research Counts*, San Francisco, CA: State of California Transportation Agency, Department of Public Works, Division of Highways, District 4, December 1969.

11. *Sixth Progress Report on Trip Ends Generation Research Counts*, San Francisco, CA: State of California Transportation Agency, Department of Public Works, Division of Highways, District 4, December 1970.

12. *Seventh Progress Report on Trip Ends Generation Research Counts*, San Francisco, CA: State of California Transportation Agency, Department of Public Works, Division of Highways, District 4, December 1971.

13. *Eighth Progress Report on Trip Ends Generation Research Counts*, San Francisco, CA: State of California Transportation Agency, Department of Public Works, Division of Highways, District 4, December 1973.

14. *Trip Generation Study*, Baltimore, MD: Maryland State Road Commission, Bureau of Traffic Planning, 1968.

15. *Trip Generation Study*, Baltimore, MD: Maryland State Road Commission, Bureau of Transportation Planning, 1970.

16. *Special Traffic Generator Study*, Report Number 1, Residential Generations, Dover, DE. State of Delaware, Department of Highways and Transportation, 1971.

17. *Special Traffic Generator Survey*, Industrial Generations, Dover, DE: State of Delaware, Department of Highways and Transportation, 1972.

18. *First Progress Report on Traffic Generators*, San Diego, CA: State of California Business and Transportation Agency, Department of Public Works, Division of Highways, District 11, December 1971.

19. *Second Progress Report on Traffic Generators*, San Diego, CA: State of California Business and Transportation Agency, Department of Public Works, Division of Highways, District 11, December 1972.

20. *Comparison of Virginia Urban Trip Generation Studies with Similar Investigations Conducted by the States of Maryland and California*, Richmond, VA: Virginia Department of Highways, Metropolitan Transportation Planning Division, July 1972.

21. *Trip Generation Rates*, Interim Technical Report 4365-4410, New York City, NY: Tri-State Regional Planning Commission, May 1973.

22. *Transportation Considerations of Regional Shopping Centers*, Interim Technical Report, New York City, NY: Tri-State Transportation Commission, November 1969.

23. *Travel to General Hospitals*, Interim Technical Report, New York City, NY: Tri-State Transportation Commission, March 1970.

24. *La Crosse Area Transportation Study*, Survey Data, La Crosse, WI: Wisconsin Department of Transportation, Division of Highways, 1970.

25. *Janesville Area Transportation Study*, Janesville, WI: Wisconsin Department of Transportation in cooperation with the United States

Department of Transportation, Federal Highway Administration, 1973.

26. *Lexington Transportation Study*, Special Generator Study, Frankfort, KY: Kentucky Department of Highways, July 1972.

27. *Trip Generating Potential of Hospitals*, Biciunas, A.E., Research News (Chicago Area Transportation Study), Volume 7, Number 4, December 31, 1965.

28. *Access and Parking Criteria for Hospitals*, Pendakur, V. Setty, and Paul O. Roer, Highway Research Record 371, 1971.

29. *Traffic Generation Study of Rest Homes and Chronic and Convalescent Homes,* Wethersfield, CT: Connecticut Department of Transportation, December 1972.

30. *Traffic and Parking Requirements of Off-Center Medical Office Building*, Shuldiner, Paul W., Donald S. Berry and James Montgomery, Jr., Highway Research Record 49, 1963.

31. *Trip Generation Equations by Zone*, Dougherty Area Regional Transportation Study, Atlanta, GA: State of Georgia Department of Transportation, February 1969.

32. *Floyd-Rome Urban Transportation Study*, Technical Report Number Six, Documentation of Model Development and Calibration, Atlanta, GA: State of Georgia Department of Transportation, July 1972.

33. *Trip Generation Summary*, Trip Generation Study of Kaiser Koulai Clinic, Study Number TG-006, Revised, Honolulu, HI: State of Hawaii Department of Transportation, Highways Division - Planning Branch.

34. *Trip Generation*, Las Cruces Area Transportation Study, City of Las Cruces, NM: New Mexico State Highway Department, 1970.

35. *Composite Report of Traffic Generation Studies*, Los Angeles, CA: City of Los Angeles, August 1969.

36. *Trip Making Characteristics*, Los Angeles, CA: City of Los Angeles, November 1972.

37. *Single Family Generation Study—Summary*, Los Angeles, CA: City of Los Angeles, June 1972.

38. *Trip Generation Rates*, unpublished report, County of Los Angeles, Los Angeles, CA, March 1973.

39. *Single Family Dwelling Unit Trip Generation Factors From Department of Traffic Studies*, Los Angeles, CA: City of Los Angeles, June 1973.

40. *Trip Generation Summary*, Honolulu, HI: State of Hawaii, Department of Transportation, Highways Division, Planning Branch, 1972.

41. "A Methodology for Determining the Traffic Impact of Regional Shopping Centers," Cohen, Daniel S., unpublished report, Washington, D.C., July 1970.

42. *Trip Generation*, Kimmel, H., S.E. Rowe, A. Rubenstein, R. Stanford, and A. Weber, Los Angeles, CA: Automobile Club of Southern California, January 1967.

43. *Traffic Characteristics of Shopping Centers*, Washington, D.C.: Metropolitan Washington Council of Governments, July 1970.

44. *Study of Eighteen Alder Community Shopping Centers*, Hollander Associates, Sidney, Baltimore, MD: The Mayor's Small Business Administration, City of Baltimore, November 1965.

45. *Characteristics of Shopping Centers*, Stoll, Walter, Chicago, IL: Chicago Area Transportation Study, State of Illinois, July 1966.

46. *Residential Locations and Shopping Patterns in Oakland County*, Cousens, Patricia D., William M. Ladd and David A. Pampu, Pontiac, Michigan, August 1966.

47. *An Approximation of Regional Shopping Center Traffic*, Buttke, Carl H., Traffic Engineering, April 1972.

48. *Trip Generation at Shopping Centers*, Miller, Forrest D., Traffic Engineering, September 1969.

49. *Urban Travel Patterns for Airports, Shopping Centers, and Industrial Plants*, Keefer, Louis E., Washington, D.C.: Highway Research Board (NCHRP Report 24), 1966.

50. *Urban Travel Patterns for Hospitals, Universities, Office Buildings and Capitols*, Keefer, Louis E., and David K. Witheford, Washington, D.C.: Highway Research Board (NCHRP Report 62), 1969.

51. *Parking and Traffic Generation—Office Buildings*, Newport Beach, CA: Herman Kimmel and Associates, September 1970.

52. *Traffic Generation and Parking Factors*, St. Paul: Barton-Aschman Associates, Inc., 1971.

53. *Summary of Special Generator Studies*, Minneapolis: Twin Cities Area Metropolitan Council, 1971.

54. *Travel Generation,* Cedar Rapids, IA: National Association of County Engineers, Action Guide Series, July 1972.

55. "Parking and Trip Generation Reports and Summaries," Volume 1, unpublished report, DeLeuw, Cather & Company, Chicago, IL.

56. *Atlantic City Urban Area Transportation Study,* Survey Data, Atlantic City, NJ: New Jersey Department of Transportation, 1973.

57. *Traffic Generator Study,* Shopko West Shopping Center, Steinhauer, J.J., Green Bay, WI: Wisconsin Department of Transportation, December 1967.

58. *Traffic Generator Study,* Arlan's Shopping Center, Sheboygan, WI: Wisconsin Department of Transportation, April 1967.

59. *Traffic Generated by Shopping Centers,* Degiorgi, Bruno R., Poughkeepsie, NY: New York State Department of Transportation, Planning Group, Region 8, 1971.

60. Palo Alto Transportation Planning Program, Data Measured at Stanford Shopping Center, Palo Alto, CA: DeLeuw, Cather & Company, 1969.

61. *Characteristics of Travel to a Regional Shopping Center,* Silver, Jacob, and Walter G. Hansen, Public Roads, December 1960.

62. Oregon State Highway Division Studies of three shopping centers at confidential locations, unpublished, Salem, OR, 1972.

63. *Report of Traffic and Engineering Investigation of Mayfair Shopping Center Driveways on S.T.H. 100 and West North Avenue,* City of Wauwatosa, WI: State Highway Commission of Wisconsin, December 1964.

64. *Shopping Center Study,* Messner, William H., Wethersfield, CT: Connecticut Highway Department, June 1968.

65. *Shopping Centers: Planning and Design for Traffic and Traffic Generation,* Harding, C.H.V., Berkeley, CA: The Institute of Transportation and Traffic Engineering, University of California, August 1960.

66. "Traffic Generator," Motorola, Inc., unpublished paper, Schaumberg, IL, 1972.

67. "Traffic Generator," William Harper Jr. College, unpublished paper, 1971.

68. *Land Use Master Plan,* Swan Island Industrial Park, Portland, OR: Planning Division, Port of Portland, 1982.

69. *Transportation Considerations of Regional Shopping Centers,* ITE Technical Council Committee 6V-A, Traffic Engineering, August 1972.

70. *Guideways for Driveway Design and Location,* Arlington, VA: Institute of Traffic Engineers, 1975.

71. *Residential Trip Generation,* Interim Technical Report, New York City, NY: Tri-State Transportation Commission, May 1971.

72. *Trip Generation Statistics,* New York City, NY: New York Metropolitan Section ITE, January 1973.

73. *Trip Generation Studies of Three Regional Shopping Centers in Washington,* Olympia, WA: Washington State Department of Highways, 1973.

74. *Industrial Park Trip Generation Study,* Wethersfield, CT: Connecticut Department of Transportation, Bureau of Planning and Research, 1972.

75. Study Number TG-005, Revised, Honolulu, HI: State of Hawaii, Department of Transportation, Highways Division, Planning Branch, 1972.

76. *Variations in Traffic Flow at Regional Shopping Centers,* Gern, Richard C., prepared for the Canadian Good Roads Association, Evanston, IL: Barton-Aschman Associates, Inc., September 1968.

77. "Various Studies of Shopping Centers," unpublished report, Simpson & Curtin, Philadelphia, PA.

78. *Special Traffic Generator Study,* Shopping Centers, Dover, DE: State of Delaware, Department of Highways and Transportation, 1972.

79. *Travel to Regional Shoping Centers,* Interim Technical Report, New York City, NY: Tri-State Transportation Commission, January 1970.

80. *Trip Generation,* Santa Ana, CA: Office of County Surveyor and Road Commissioner, Orange County, October 1972.

81. "Land Use and Traffic Generation Characteristics of Rural Highway Interchanges," unpublished data, Kuhn, Herman A.J., University of Wisconsin, Madison, Wisconsin, 1967.

82. *Trip Generation by Land Use,* Tempe, AZ: Maricopa Association of Governments, April 1974.

83. *Parking and Access at General Hospitals,* Kanaan, George E., Westport, CT: Eno Foundation for Transportation, Inc., 1973.

84. *Special Traffic Generator Study,* Report Number 1, Residential Generations, Dover, DE, Department of Highways and Transportation, 1974, Revision 2.

85. *Special Traffic Generator Study,* Report Number 2, Industrial Generations, Dover, DE, Department of Highways and Transportation, 1973.

86. *Special Traffic Generator Study,* Report Number 3, Education Facilities Generations, Dover, DE, Department of Highways and Transportation, 1976.

87. *Special Traffic Generator Study,* Report Number 4, Commercial Generations, Dover, DE, Department of Highways and Transportation, 1975.

88. *Ninth Progress Report on Trip Ends Generation Research Counts,* San Francisco, CA: State of California, Department of Transportation, District 4, July 1974.

89. *Tenth Progress Report on Trip Ends Generation Research Counts,* San Francisco, CA: State of California, Department of Transportation, District 4, July 1975.

90. *Eleventh Progress Report on Trip Ends Generation Research Counts,* San Francisco, CA: State of California, Department of Transportation, District 4, July 1976.

91. *Trip Generation Study of Various Land Uses,* Wethersfield, CT: Connecticut Department of Transportation, June 1974.

92. *Trip Generation Study of Various Land Uses,* Supplement A, Zevin, Israel, Wethersfield, CT: Connecticut Department of Transportation, March 1975.

93. *Trip Generation Study of Regional Shopping Centers,* Mid-Ohio Regional Planning Commission, Columbus, OH, November 1977.

94. *Re-Evaluation of Trip Generation Study of Condominium Developments in the Columbus Metropolitan Area,* Mid-Ohio Regional Planning Commission, Columbus, OH, June 1976.

95. Unpublished trip generation studies, Buttke, Carl H., Portland, OR, 1977.

96. Unpublished trip generation studies, The Municipality of Metropolitan Toronto, Department of Roads and Traffic, Toronto, Ontario, Canada, 1978.

97. *Trip Generation Data,* ITE District 7, Alberta, Canada, 1978.

98. Unpublished trip generation studies, Paul C. Box and Associates, Skokie, IL, 1981.

99. Unpublished trip generation studies, Hensley-Schmidt, Inc., Consultants, Chattanooga, TN, 1981.

100. *Traffic Generators,* San Diego Association of Governments, 1979 to 1981.

101. Unpublished trip generation studies, Transportation Planning and Engineering, Inc., Bellevue, WA, 1981.

102. Unpublished trip generation studies, Tippetts-Abbett- McCarthy-Stratton, Boston, MA, 1981.

103. Unpublished trip generation studies, Denver Regional Council of Governments, Denver, CO, 1980.

104. Unpublished trip generation studies, Bather, Ringrose, Wolsfeld, Jarvis, Gardner, Inc., Minneapolis, MN, 1981.

105. Unpublished trip generation studies, Barton-Aschman Associates, Inc., Washington, D.C., 1981.

106. Unpublished trip generation studies, Grigg, Glenn M., Cupertino, CA, 1980.

107. *Access Study,* General Telephone Headquarters Expansion, Entranco Engineers, Bellevue, WA, 1979.

108. *Traffic Generation Survey,* Gloucester County Planning Department, Clayton, NJ, 1979.

109. Unpublished trip generation studies, Schimpeler, Corradino Associates, Louisville, KY, 1980.

110. *Special Land Use Trip Generation in Virginia,* Arnold, E.D., Jr., Virginia Highway and Transportation Research Council, Charlottesville, VA, 1981.

111. Unpublished trip generation studies, Metcalf, Gary W., Overland Park, KS, 1980.

112. Unpublished trip generation study, Virginia Polytechnic Institute and State University, 1980.

113. *Twelfth Progress Report on Trip Ends Generation Counts,* San Francisco, CA: State of California, Department of Transportation, District 4, December 1979.

114. *Thirteenth Progress Report on Trip Ends Generation Counts,* San Francisco, CA: State of California, Department of Transportation, District 4, December 1980.

115. Unpublished trip generation study, Herp, Donald J., Phoenix, AZ, 1980.

116. *Trip Generation Data,* ITE District 7, Toronto Section, Toronto, Ontario, Canada, 1978.

117. *Trip Generation Study,* Nash, Bruce, Chico, CA, 1978.

118. Unpublished trip generation measurements, Brown, Christopher, Seattle, WA, 1981.

119. Unpublished trip generation studies, Pleyte, Allan P., Milwaukee, WI, 1980.

120. *Generation Studies,* The Center for Urban Transportation Studies, University of Wisconsin-Milwaukee, Trip Milwaukee, WI, 1980.

121. *Marine Terminal Traffic Generation Manual,* Wilbur Smith and Associates, for the Metropolitan Transportation Commission and the Bay Conservation and Development Commission, San Francisco, CA, 1980.

122. *Trip Ends Generation Study,* Pursell, Gary, Chico, CA, 1978.

123. Unpublished trip generation studies, Wilsey & Ham, Inc., Seattle, WA, 1981.

124. *Traffic Generation Characteristics: Florida Shopping Centers,* Byrne, A.S., ITE Technical Notes, Fall 1975.

125. Unpublished trip generation studies, City of Milwaukee, Bureau of Traffic Engineering, Milwaukee, WI, 1980.

126. *Fourteenth Progress Report on Trip Ends Generation Research Counts,* California Department of Transportation, District 4, San Francisco, CA, July 1982.

152. *Special Land Use Trip Generation at Special Sites,* Virginia Highway and Transportation Research Council, January 1984.

153. "Trip Generation Rates," Unpublished, West Virginia Department of Transportation.

154. Unpublished trip generation studies, New York Department of Transportation, Region 1, Albany, NY, 1984.

155. Unpublished trip generation studies, Carl H. Buttke, Inc., Portland, OR, Metropolitan Area, 1980-1984.

156. Unpublished trip generation studies, Vanasse Hangen Associates, Inc., Boston, MA, 1982.

157. Unpublished trip generation studies, Thomas S. Montgomery & Associates, CA, 1983.

158. Unpublished trip generation studies, Crommelin-Pringle & Associates, Los Angeles, CA, 1974.

159. Unpublished trip generation studies, Wes Guckert, Parkton, MD, 1983.

160. Unpublished trip generation studies, Segal DiSarcina Associates, Boston, MA, 1982.

161. Unpublished trip generation studies, C.E. Maguire, Inc., New Britain, CT, 1984.

162. Unpublished trip generation studies, New York Department of Transportation, Albany, NY, 1984.

163. Unpublished trip generation studies, ITE Intermountain Section, Billings, MT, 1982.

164. Unpublished trip generation studies, Sear-Brown Associates, P.C., Rochester, NY, 1985.

165. Unpublished trip generation studies, BRW, Inc., Minneapolis, MN, 1984.

166. Unpublished trip generation studies, Street Traffic Studies, Ltd., Baltimore, MD, June 1984.

167. Unpublished trip generation studies, California State University, Chico, CA, April 1984.

168–170. *Trip Generation at Special Sites, Final Report,* Virginia Highway and Transportation Research Council, Commonwealth of Virginia, February 1984.

171. Unpublished trip generation studies, City of Lakewood, CO, July 1985.

172–173. Unpublished trip generation studies, Barton-Aschman Associates, Houston, TX, 1979-1985.

174. Unpublished trip generation studies, Leigh, Scott & Cleary, Inc., Colorado Springs, CO, January 1985.

175. Unpublished trip generation studies, Traffic Engineering and highway Safety, Westchester County, White Plains, NY, 1984.

176. Unpublished trip generation studies, Department of Public Works, City of Lakewood, CO, March 1985.

177. Unpublished trip generation studies, Barge, Waggoner, Sumner and Cannon, Nashville, TN, 1984-1985.

178. Unpublished trip generation studies, Horner and Canter Associates, Medford, NJ, 1984-1985.

179. Unpublished trip generation studies, City of Corvallis Utility and Engineering Services, Corvallis, OR, 1985.

180–181. Unpublished trip generation studies, City of Milwaukee, WI, Department of Public Works, 1983.

182. Unpublished trip generation studies, Barton-Aschman Associates, Inc., Houston, TX, March 1979.

183. Unpublished trip generation studies, Dallas, TX, Texas Department of Transportation, 1985.

184. *Nassau County Trip Generation Report,* Office and Industrial Uses, Nassau County Planning Commission, New York Metropolitan Transportation Council, June 1986.

185–186. *Westchester County Trip Generation Study,* Draft Final Report, New York Metropolitan Transportation Council, New York, NY, May 1985.

187. *Suffolk County Trip Generation Study,* New York Metropolitan Transportation Council, Suffolk County Planning Department, August 1985.

188. Unpublished trip generation studies, RBA Group, Atlanta, GA, October 1984.

189. Unpublished trip generation studies, Orth-Rodgers & Associates, Inc., Philadelphia, PA, August 1985.

190. Unpublished trip generation studies, Des Moines Traffic and Transportation Department, Des Moines, IA, 1986.

191. Unpublished trip generation studies, Travers Associates, Inc., Clifton, NJ, 1985.

192–203. Unpublished trip generation studies, Raymond Keyes Associates, P.C., Elmsford, NY, 1984.

204. *Trip Generation Rates for Multiple Family Residential Developments and Neighborhood Shopping Centers in the Chicago Area,* Technical Memorandum 83-01, Civgin, Mehmet, Chicago Area Transportation Study, December 1982.

205. *Fifteenth Progress Report on Trip Ends Generation Research Counts,* Chang, Herman and Andrzej Wolny, State of California, Department of Transportation, District 4, Transportation Studies Branch, San Francisco, CA, December 1983.

206. *Traffic Analysis —Wild Waters,* Shandro, Peter, Ada County Highway Department, Boise, ID, October 25, 1986.

207. Unpublished trip generation studies, Crawford, Bunte, Brammeier, St. Louis, MO, January 1982.

208. *Trip Generation Study,* Kinder Care Learning Centers, Street Traffic Studies, Ltd., Gaithersburg, MD, May 1984.

209. *Trip Generation Analysis,* Christiana Medical Offices Project, Ryan, Timothy A., Kidde Consultants, Inc., Baltimore, MD, February 1984.

210. Unpublished trip generation studies, Fitzpatrick, Douglas R., Fitzpatrick-Llewellyn, Inc., Essex Junction, VT, June 8, 1984.

211. *The Brandermill PUD Traffic Generation Study,* Technical Report, JHK and Associates, Alexandria, VA, June 1984.

212–214. Unpublished trip generation studies, San Diego Association of Governments, San Diego, CA, August 1986.

215. "Movie Theater Trip Generation Rates," Baumgaertner, William E., *ITE Journal,* Washington, D.C., June 1985.

216. *Trip Generation Rates for New Types of Generators,* The RBA Group, Voorhies, Kenneth O., Atlanta, GA, March 1986.

217. Unpublished trip generation studies, City of Troy, MI, Beaubien, Richard F., March 31, 1986.

218. Unpublished trip generation studies, Montgomery County Government, Rockville, MD, October 30, 1984.

219. *Video Arcade Traffic and Parking,* Reynolds/ Russillo, 1983.

220. Unpublished trip generation studies, State of Vermont, Agency of Transport, Montpelier, VT, 1988.

221. Unpublished trip generation studies, Transportation Department, City of Edmonton, Alberta, Canada, March 1988.

222–235. Miscellaneous unpublished trip generation studies

236. Unpublished trip generation study, Clough Harbour and Associates, Albany, NY, September 1988.

237. Unpublished trip generation studies, Works Department, Traffic Division, Etobicoke, Ontario, Canada, November 1982.

238. Unpublished trip generation studies, Los Angeles County of Public Works, Los Angeles, CA, 1989.

239. Unpublished trip generation study, Orth-Rodgers and Associates, Raritan, NJ, June 1988.

240. Unpublished trip generation studies, County of San Louis Obispo, CA, March 1989.

241. Unpublished trip generation studies, McMahon Associates, Inc., Willow Grove, PA, September 1987.

242-243. Miscellaneous unpublished trip generation studies

244. Unpublished trip generation studies, McMahon Associates, Inc., Willow Grove, PA, September 1987.

245. Unpublished trip generation studies, Frederick P. Clark Associates, Consultants, Southport, CT, July 1987.

246. Unpublished trip generation study, Yuma Metropolitan Planning Organization, Yuma, AZ, January 1989.

247. Unpublished trip generation studies, Maguire Group, Inc., CT, April 1989.

248. Unpublished trip generation study, ITE Student Chapter, Purdue University, West Lafayette, IN, March 1989.

249. Miscellaneous unpublished trip generation studies.

250. Unpublished trip generation study, Nashua Regional Planning Commission, Nashua, NH, May 1989.

251. Unpublished trip generation studies, Orth-Rodgers and Associates, Inc., PA, October 1988.

252. Unpublished trip generation study, A&F Engineering, Inc., Indianapolis, IN.

253. Unpublished trip generation studies, McMahon Associates, Inc., Willow Grove, PA, 1980s.

254. Unpublished trip generation studies, McMahon Associates, Inc., Willow Grove, PA, 1987.

255. Unpublished trip generation studies, Weston Pringle and Associates, Fullerton, CA, August 1988.

256. Unpublished trip generation studies, Traffic Planning & Design Inc., Oaks, PA, 1989.

257. Unpublished trip generation study, Port Authority of New York and New Jersey, New York, NY, October 1988.

258. Unpublished trip generation studies, Vollmer Associates, North Haledon, NJ, 1989.

259. Unpublished trip generation studies, JBM and Associates Traffic Study, Submitted by the City of Overland Park, KS, September 1988.

260. Unpublished trip generation studies, A&F Engineering Co., Indianapolis, IN.

261. Unpublished trip generation study, Travers Associates, Inc., Ridgewood, NJ, December 1988.

262. Unpublished trip generation studies, Andrews and Clark, Inc., Long Island, NY, June 1987.

263. Unpublished trip generation study, Detroit DOT, Detroit, MI, January 1989.

264. Unpublished trip generation studies, Engineering Department, County of San Louis Obispo, CA, September 1988.

265. Unpublished trip generation study, Barkan and Mess Associates, Inc., Clinton, CT, August 1986.

266. Unpublished trip generation studies, Orth-Rodgers and Associates, Inc., Bridgewater, NJ, September/October 1989.

267. Unpublished trip generation studies, DSA Group, Inc., Bradenton, FL, June 1988.

268. Unpublished trip generation study, BRW, Inc., Bloomington, MN, June 1988.

269. Unpublished trip generation studies, Traffic Management Division, City of Oklahoma City, OK, April 1988.

270. Unpublished trip generation studies, Glatting Lopez Kercher Anglin, Orlando, FL, August 1989.

271. Unpublished trip generation studies, Nolte and Associates, Santa Cruz County, FL, October 1989.

272. Unpublished trip generation study, Department of Public Works, City of Ceres, CA, June 1989.

273. San Diego Traffic Generators, San Diego Association of Governments, San Diego, CA, September 1989.

274. Gas/convenience store trip generation study, State of Florida DOT, July 1989.

275. Unpublished trip generation studies, City of Parma, OH, March 1981.

276. "Trip Characteristics of Convenience Markets With Gas Pumps," Montana Technical Committee of the Intermountain Section of ITE, *ITE Journal*, July 1987.

277. *San Diego Traffic Generators*, San Diego Association of Governments, San Diego, CA, June 1987.

278. Unpublished trip generation studies, Parking and Traffic Department, City of Modesto, CA.

279. Unpublished trip generation studies, City of Overland Park, KS, August 1981.

280. Unpublished trip generation studies, James T. Rapoli Consulting, Poughkeepsie, NY, 1985.

281. Unpublished trip generation study, Sear-Brown Associates, P.C., Rochester, NY, September 1986.

282. Unpublished trip generation studies, Tom R. Lancaster, P.E., Portland, OR, January 1987.

283. Unpublished trip generation studies, Carl H. Buttke, Inc., Portland, OR, April 1988.

284. Unpublished trip generation study, ASL Consulting Engineers, Inc., Los Angeles, CA, August 1987.

285. "Trip Generation Characteristics of Air Force Bases," Kim Eric Hazarvartian, *ITE Journal*, October 1988.

286. Unpublished trip generation study, The Maguire Group, CT, February 1987.

287. *San Diego Traffic Generators*, San Diego Association of Governments, San Diego, CA, November 1987.

288. Unpublished trip generation studies, New England Section, ITE, 1987-1989.

289. Unpublished trip generation study, Transportation/Traffic Division, Department of Engineering Services, City of Camarillo, CA, August 1988.

290. Unpublished trip generation study, Keith and Schnars, Fort Lauderdale, FL, February 1987.

291. Unpublished trip generation studies, Daubert Engineering Corp., Colorado Springs, CO, March 1987.

292. Miscellaneous unpublished trip generation studies.

293. Unpublished trip generation studies, Maguire Group, Inc., New Britain, CT, November 1987.

294. *Trip Generation Rates for Drive-In/Fast-Food Restaurant and Medical Office Buildings in the OKI Region*, Ohio-Kentucky-Indiana Regional Council of Governments, Cincinnati, OH, June 1987.

295. Unpublished trip generation studies, Clough, Harbour and Associates, Albany, NY, September 1986.

296. Unpublished trip generation study, ITE Student Chapter at Purdue University, West Lafayette, IN, April 1987.

297. Unpublished trip generation studies, PHR & A, Fairfax, VA, 1988.

298. Unpublished trip generation studies, TJKM Transportation Consultants, Pleasanton, CA, October/November 1988.

299. Unpublished trip generation studies, The Cafaro Company, Youngstown, OH, November 1988.

300–301. *Indianapolis/Marion County Site Trip Generation Counts*, Barton-Aschman Associates, Indianapolis, IN, 1989.

302–306. *Travel Characteristics at Large-Scale Suburban Activity Centers*, NCHRP Report 323, Transportation Research Board, National Research Council, 1987-1988.

307–318. Miscellaneous unpublished trip generation studies.

319-325. *Montgomery County Trip Generation Rate Study*, Maryland-National Capital Park and Planning Commission, Silver Spring, MD, 1986.

326. Unpublished trip generation study, Citrus County Department of Development Services, Lecanto, FL, February 1990.

327. *Trip Generation From Suburban Office Buildings in New Jersey,* Delaware Valley Regional Planning Commission, Philadelphia, PA, 1988-1989.

328. Unpublished trip generation studies, J.W. Buckholz Traffic Engineering, Inc., Jacksonville, FL, September/October 1990.

329. Unpublished trip generation study, H.W. Moore Associates, Inc., Consulting Engineers, Boston, MA, March 1990.

330. Unpublished trip generation studies, Maguire Group, Inc., CT, 1990.

331. *Trip Generation Analysis for High Cube Warehouses,* Fehr and Peters Associates, City of Livermore, CA, June 1989.

332. *Sixteenth Progress Report on Trip Ends Generation Research Counts,* California Department of Transportation, District 4, December 1986.

333. Unpublished trip generation study, Barton-Aschman Associates, Inc., Columbus, OH, 1990.

334. Unpublished trip generation studies, Orth-Rodgers and Associates, Inc., Bridgewater, NJ, 1989.

335. Unpublished trip generation studies, Michael Monteleone, Chapel Hill, NC, 1990.

336. Unpublished trip generation studies, Metro Traffic and Parking, Nashville, TN, 1991.

337. Unpublished trip generation studies, City of Chattanooga, Chattanooga, TN, 1990.

338. Unpublished trip generation studies, Kentuckiana Regional Planning and Development Agency/Jefferson County Public Works and Transportation Division, Louisville, KY, 1993.

339. Unpublished trip generation studies, Travers Assoc., Inc., NJ, 1991.

340. Unpublished trip generation studies, Benshoof and Assoc., Inc., Edina, MN, 1993.

341. Unpublished trip generation studies, Traffic Planning and Design, Altamonte Springs, FL, 1992.

342. Unpublished trip generation studies, MWCOG, Washington, DC, 1989.

343. Unpublished trip generation studies, Citrus County Dev. Serv. Dept., Lecanto, FL, 1992.

344. Unpublished trip generation studies, Central Transportation Planning Staff, Boston, MA, 1992.

345. Unpublished trip generation studies, The Traffic Group, Inc., Towson, MD, 1992.

346. Unpublished trip generation studies, Muncaster Engr. & Computer Applications, Charlottesville, VA, 1990.

347-348. Mobil National Traffic Study, KHR Associates, Irvine, CA, 1992.

349. Unpublished trip generation studies, VHB, Inc., Watertown, MA, 1992.

350. Kentuckiana Regional Planning and Development Agency/Jefferson County Public Works and Transportation Division, Louisville, KY, 1993.

351. Unpublished trip generation studies, Palm Beach Co. Traf. Div., West Palm Beach, FL, 1989.

352. Unpublished trip generation studies, Tipton Assoc., Inc., Orlando, FL, 1989.

353. *Traffic Generation Study for Wal-Mart Stores,* Robert D. Vanasse & Assoc., Inc., Andover, MA, 1994.

354. *Trip generation studies for Wal-Mart Supercenters,* Peters & Assoc., Little Rock, AR, 1994.

355. *Development of Montgomery County Trip Generation Rates,* Maryland-National Capital Park and Planning Commission, Montgomery County, October 1993.

356. *Trip Generation Study,* Street Smarts, Atlanta, GA, 1990.

357-359. *Rapid City MPO, Trip Generation Rates,* City of Rapid City, SD, 1995.

360. *Trip Generation Study,* Delaware Valley Regional Planning Commission, Philadelphia, PA, 1989-1990.

361. *Trip Generation—Golf Driving Range,* Bruce Campbell & Associates, Inc., Boston, MA, 1993.

362. *Trip Generation Information for Quick Lubrication Shops in Vancouver, WA,* Kittelson & Associates, Inc., Portland, OR, October 1995.

363. *Trip Generation Study,* MCV Associates, Inc., McLean, VA, August 1994.

364. *Trip Generation Study,* Robert L. Morris, Inc., Bethesda, MD, August 1994.

365. *Trip Generation Study,* The Sear-Brown Group, Rochester, NY, 1991-1995.

366. *Trip Generation Study,* Inland Pacific Engineering Company, Spokane, WA, May 1995.

367. *Trip Generation Study,* Eschbacher & Associates, Syosset, NY, March 1996.

368. *Trip Generation Study,* Brigham Young University, Provo, UT, March 1996.

369. *Trip Generation Study,* Transportation Concepts, Clifton Park, NY, April 1996.

370. *Trip Generation Study,* TDA, Inc., Seattle, WA.

371. *Trip Generation Study,* Virginia Transportation Research Council, Charlottesville, VA, 1996.

372. *Trip Generation Study,* Grove Miller Engineering, Inc., Harrisburg, PA, 1992.

373. *Trip Generation Study,* Tulare County Association of Governments, Visalia, CA, November 1993.

374. *Trip Generation Study,* Transportation Engineers, Inc., Fullerton, CA, 1990.

375. *Trip Generation Study,* David Plummer & Associates, Inc., Coral Gables, FL, July 1992.

376. *An Informational Trip Generation Report—'Big Box Users' and 'Category Killers' for Power Retail Centers,* The Traffic Group, Towson, MD, 1993.

377. *Trip Generation and Parking Generation Study,* Optimum Environment, Issaquah, WA, 1991.

378. *Trip Generation Study,* Buckhurst Fish Hutton Katz & Jacquemart, Inc., New York, NY, 1990-1991.

379. *Trip Generation Study,* JW Buckholz Traffic Engineering, Inc., Jacksonville, FL, September 1991.

380. *Factory Outlet Center Trip Generation Study,* Associated Transportation Engineers, Santa Barbara, CA, 1991-1996.

381. *Resort Hotel Traffic Study,* Barton-Aschman Associates, Inc., Dallas, TX, 1986.

382. *Trip Generation Study,* Barr, Dunlop & Associates, Inc., Tallahassee, FL, 1995.

383. *Trip Generation Study,* Barakos-Landino Design Group, Hamden, CT, 1995.

384. *Trip Generation Study,* Benshoof & Associates, Inc., Edina, MN, 1995.

385. *Trip Generation Study,* Benshoof & Associates, Inc., Edina, MN, 1992.

386. *Free-Standing Retail Establishment Trip Generation Study,* Bergmann Associates, Rochester, NY, 1994.

387. *Trip Generation Study,* Brigham Young University, Provo, UT, March 1996.

388. *Trip Generation Study,* Langley and McDonald, Williamsburg, VA, May 1990.

389. *Trip Generation Study,* Charlotte Department of Transportation, Charlotte, NC, February 1995.

390. *Trip Generation Study,* Creative Transportation Solutions, Burnaby, B.C.,Canada, 1994-1995.

391. *Trip Generation Study,* Creighton Manning, Inc., Delmar, NY, December 1994.

392. *Trip Generation Study,* Cupertino, CA, 1993-1995.

393. *Trip Generation Study,* DJK Associates, Inc., Arlington, MA, May 1991.

394. *Trip Generation Study,* D.J. Parrone & Associates, Penfield, NY, May 1993.

395. *Trip Generation Study,* David Evans and Associates, Inc., Portland, OR, November 1991.

396. *Trip Generation Study,* City of Farmington, Farmington, NM.

397. *Trip Generation Study,* Horner & Canter Associates, Medford, NJ, 1991-1994.

398. *Trip Generation Study,* Glatting Lopez Kercher Anglin, Orlando, FL, 1990-1991.

399. *Trip Generation Study,* Grove Miller Engineering, Inc., Harrisburg, PA, January 1992.

400. *Trip Generation Study,* I.K. Chann Associates, Wilton, CT, February 1996.

401. *Trip Generation Study,* Inland Pacific Engineering Company, Spokane, WA, February 1996.

402. *Trip Generation Study,* Inland Engineering Corporation, Victorville, CA, May 1995.

403. *Traffic & Circulation Study for Proposed Mini Storage,* Transportation Engineers, Inc., Fullerton, CA, November 1993.

404. *Trip Generation Study,* Kentuckiana Regional Planning and Development Agency, Louisville, KY, 1993.

405. *Trip Generation Study,* Mackenzie Engineering, Inc., Portland, OR.

406. *Trip Generation Study,* CE Maguire, Inc., New Britain, CT, 1986-1994.

407. *County of Morris 1992 Trip Generation Study,* Morris County, NJ.

408. *Trip Generation Study,* Meyer, Mohaddes Associates, Inc., San Mateo, CA, November 1995.

409. *Trip Generation Study,* North Carolina Department of Transportation, Raleigh, NC, 1993.

410. *Trip Generation Study,* Town of Oro Valley, Oro Valley, AZ, 1993.

411. *Trip Generation Study,* Orth-Rodgers & Associates, Inc., Bridgewater, NJ, 1987-1990.

412. *Trip Generation Study,* Orth-Rodgers & Associates, Inc., Bridgewater, NJ, 1991.

413. *Trip Generation Study,* City of Overland Park, Overland Park, KS, April 1991.

414. *Trip Generation Study,* Paul C. Box & Associates, Inc., Skokie, IL, 1987-1991.

415. *Trip Generation Study,* Prosser, Hallock & Kristoff, Inc., Jacksonville, FL, January 1994.

416. *Traffic Generation Study,* Robert D. Vanasse & Associates, Inc., Andover, MA, March 1993.

417. *Trip Generation Study,* TRC Raymond Keyes Associates, Tarrytown, NY, 1994-1995.

418. Unpublished trip generation studies, Barton-Aschman Associates, Inc., San Jose, CA, 1987-1995.

419. *Trip Generation Study,* Balloffet & Associates, Inc., Denver, CO, October 1995.

420. *Trip Generation Study,* State of Vermont Agency of Transportation, Montpelier, VT, December 1990.

421. *Trip Generation Study,* Schoor DePalma, Manalapan, NJ, 1993-1996.

422-424. *Trip Generation Study,* DKS Associates, Portland, OR, 1991-1996.

425. *Trip Generation Study,* Transportation Planning & Engineering, Inc., Bellevue, WA, 1991-1992.

426. *Trip Generation Study,* Tim Miller Associates, Inc., Cold Spring, NY, September 1992.

427. *Trip Generation Study,* Area Plan Commission of Tippecanoe Co., Lafayette, IN, December 1995.

428. *Trip Generation Study,* Travers Associates, Inc., NJ, 1990-1994.

429. *Trip Generation Study,* Vollmer Associates, Rochelle Park, NJ, December 1993.

430. *Trip Generation Study,* Western Planning & Research, Inc., Auburn, CA, 1996.

431. *Trip Generation Study,* University of Tennessee Transportation Center, Knoxville, TN, 1995.

432. *District IV Trip Generation Study,* University of Wisconsin-Platteville Student Chapter, Platteville, WI, 1994-1995.

433. *Trip Generation Study,* University of Hawaii at Manoa, Honolulu, HA, November 1995.

434. *Trip Generation Study,* University of Arkansas, Fayetteville, AR, September 1995.

435. *Supplement to San Diego Traffic Generators,* San Diego Association of Governments, San Diego, CA, 1991-1995.

436-439. *Trip Generation Studies,* Traffic Planning and Design, Maitland, FL, 1991-1996.

440-441. *Trip Generation Studies,* Associated Transportation Engineers, Santa Barbara, CA.

442. *Trip Generation Study,* Sprinkle Consulting Engineering, Lutz, FL, 1990-1993.

443-445. *Trip Generation Surveys,* DKS Associates, OR, 1991-1996.

446. *International Council of Shopping Centers Trip Generation Study,* Raymond Keyes Associates Inc., Tarrytown, NY, 1994.

447. *Trip Generation Studies,* Schoor DePalma, Manalapan, NJ, 1995.

448. *Trip Generation Studies,* Connecticut Department of Transportation, Newington, CT, 1996.

449. *Lumber Store Trip Generation Analysis,* JW Buckholz Traffic Engineering Inc., Jacksonville, FL, 1992.

APPENDIX B:
DATA COLLECTION FORM
AND USER COMMENTS FORM

Institute of Transportation Engineers

Trip Generation Data Form (*Part 1*)

Land Use/Building Type:[1]		ITE Land Use Code:
Source:		Source No. (by ITE):
Name of Development:		Day of the Week:
City:	State/Province:	Zip/Postal Code:	Day:	Month:	Year:
Country:		Metropolitan Area:

1. For fast-food land use, please specify if hamburger- or nonhamburger-based.

Location Within Area:

- ☐ (1) CBD
- ☐ (2) Urban (Non-CBD)
- ☐ (3) Suburban (Non-CBD)
- ☐ (4) Suburban CBD
- ☐ (5) Rural
- ☐ (6) Freeway Interchange Area (Rural)
- ☐ (7) Not Given

Detailed Description of Development:[3]

Independent Variable: (include data for as many as possible)[2]

	Actual	Estimated		Actual	Estimated
_____ (1) Employees (#)	☐	☐	_____ (10) Parking Spaces (#)	☐	☐
_____ (2) Persons (#)	☐	☐	_____ (11) Occupied Beds (#)	☐	☐
_____ (3) Units (#)	☐	☐	_____ (12) Seats (#)	☐	☐
_____ (4) Occupied Units (#)	☐	☐	_____ (13) Servicing Positions/Vehicle Fueling	☐	☐
_____ (5) Building Area (gross sq. ft.)	☐	☐	_____ Positions		
_____ (% of development occupied _____)			_____ (14) Shopping Center % Out-parcels/pads	☐	☐
_____ (6) Net Rentable Area (sq. ft.)	☐	☐	_____ (15) AM Peak Hour Volume of Adjacent Street Traffic	☐	☐
_____ (7) Gross Leasable Area (sq. ft.)	☐	☐	_____ (16) PM Peak Hour Volume of Adjacent Street Traffic	☐	☐
_____ (8) Occupied Gross Leasable Area (sq. ft.)	☐	☐	_____ (17) Other _____	☐	☐
_____ (9) Acres	☐	☐	_____ (18) Other _____	☐	☐

2. Definitions for several independent variables can be found in the *Trip Generation User's Guide.*

3. Please provide all pertinent information that helps to describe the subject project. If necessary, attach a detailed report.

Other Data:

Vehicle Occupancy (#)

_____ AM _____ PM _____ 24-hour %

Percent by Transit:

_____ AM % _____ PM % _____ 24-hour %

Percent by Carpool/Vanpool:

_____ AM % _____ PM % _____ 24-hour %

Full-time Employees by Shift:

	Start Time	End Time	
First Shift:	_____	_____	_____ Employees (#)
Second Shift:	_____	_____	_____ Employees (#)
Third Shift:	_____	_____	_____ Employees (#)

Parking Cost on Site: Hourly _____ Daily _____

Transportation Demand Management (TDM) Information:

At the time of this study, was there a TDM program (that may have impacted the trip generation characteristics of this site) under way?

☐ No

☐ Yes (If yes, please check appropriate box/boxes, describe the nature of this TDM program(s) and provide a source for any studies that may help quantify this impact. Attach additional sheets if necessary)

- ☐ (1) Transit Service
- ☐ (2) Carpool Programs
- ☐ (3) Vanpool Programs
- ☐ (4) Bicycle/Pedestrian Facilities and Site Improvements
- ☐ (5) Employer Support Measures
- ☐ (6) Preferential HOV Treatments
- ☐ (7) Transit and Ridesharing Incentives
- ☐ (8) Parking Supply and Pricing Management
- ☐ (9) Tolls and Congestion Pricing
- ☐ (10) Variable Work Hours/Compressed Work Weeks
- ☐ (11) Telecommuting
- ☐ (12) Other _____

Please Complete Form on Other Side

ITE Institute of Transportation Engineers

Trip Generation Data Form (*Part 2*)

(All = All Vehicles Counted; Trucks = Heavy Duty Trucks & Buses)

Summary of Driveway Volumes

	Average Weekday (M-F)						Saturday						Sunday					
	Enter		Exit		Total		Enter		Exit		Total		Enter		Exit		Total	
	All	Trucks	All	Trucks	All	Trucks	All	Trucks	All	Trucks	All	Trucks	All	Trucks	All	Trucks	All	Trucks
24-Hour Volume																		
A.M. Peak Hour of Adjacent[1] Street Traffic (7–9) Time:																		
P.M. Peak Hour of Adjacent Street Traffic (4–6) Time:																		
A.M. Peak Hour: Generator[2] Time:																		
P.M. Peak Hour: Generator Time:																		
No. of Days Counted																		

1, 2. Please refer to the *Trip Generation User's Guide* for a definition of the terms.

Detailed Driveway Volumes—Average Weekday (M-F)

A.M. Period	Enter		Exit		Total		Mid-Day Period	Enter		Exit		Total		P.M. Period	Enter		Exit		Total	
	All	Trucks	All	Trucks	All	Trucks		All	Trucks	All	Trucks	All	Trucks		All	Trucks	All	Trucks	All	Trucks
6:00-6:15							11:00-11:15							3:00-3:15						
6:15-6:30							11:15-11:30							3:15-3:30						
6:30-6:45							11:30-11:45							3:30-3:45						
6:45-7:00							11:45-12:00							3:45-4:00						
7:00-7:15							12:00-12:15							4:00-4:15						
7:15-7:30							12:15-12:30							4:15-4:30						
7:30-7:45							12:30-12:45							4:30-4:45						
7:45-8:00							12:45-1:00							4:45-5:00						
8:00-8:15							1:00-1:15							5:00-5:15						
8:15-8:30							1:15-1:30							5:15-5:30						
8:30-8:45							1:30-1:45							5:30-5:45						
8:45-9:00							1:45-2:00							5:45-6:00						
9:00-9:15														6:00-6:15						
9:15-9:30														6:15-6:30						

☐ Please attach any additional site information or comments regarding special site characteristics, if available.

☐ Check if additional information is attached.

Survey conducted by: Name: _____

Organization: _____

Address: _____

City/State/Zip: _____

Telephone #: _____ Fax #: _____ E-mail: _____

Please *return to:* Institute of Transportation Engineers
Programs & Services Department
525 School Street, SW, Suite 410
Washington, DC 20024-2797 USA
Telephone: +1 202/554-8050
FAX: +1 202/863-5486
ITE on the Web: http://www.ite.org

ITE Institute of Transportation Engineers

Trip Generation Data Form (*Part I*)

Land Use/Building Type:[1]		ITE Land Use Code:
Source:		Source No. (by ITE):
Name of Development:		Day of the Week:
City:	State/Province:	Zip/Postal Code:	Day:	Month:	Year:
Country:		Metropolitan Area:

1. For fast-food land use, please specify if hamburger- or nonhamburger-based.

Detailed Description of Development:[3]

Location Within Area:

- ☐ (1) CBD
- ☐ (2) Urban (Non-CBD)
- ☐ (3) Suburban (Non-CBD)
- ☐ (4) Suburban CBD
- ☐ (5) Rural
- ☐ (6) Freeway Interchange Area (Rural)
- ☐ (7) Not Given

Independent Variable: *(include data for as many as possible)[2]*

	Actual	Estimated		Actual	Estimated
___ (1) Employees (#)	☐	☐	___ (10) Parking Spaces (#)	☐	☐
___ (2) Persons (#)	☐	☐	___ (11) Occupied Beds (#)	☐	☐
___ (3) Units (#)	☐	☐	___ (12) Seats (#)	☐	☐
___ (4) Occupied Units (#)	☐	☐	___ (13) Servicing Positions/Vehicle Fueling	☐	☐
___ (5) Building Area (gross sq. ft.)	☐	☐	___ Positions		
___ (% of development occupied _____)			___ (14) Shopping Center % Out-parcels/pads	☐	☐
___ (6) Net Rentable Area (sq. ft.)	☐	☐	___ (15) AM Peak Hour Volume of Adjacent Street Traffic	☐	☐
___ (7) Gross Leasable Area (sq. ft.)	☐	☐	___ (16) PM Peak Hour Volume of Adjacent Street Traffic	☐	☐
___ (8) Occupied Gross Leasable Area (sq. ft.)	☐	☐	___ (17) Other _____	☐	☐
___ (9) Acres	☐	☐	___ (18) Other _____	☐	☐

2. Definitions for several independent variables can be found in the *Trip Generation User's Guide.*

3. Please provide all pertinent information that helps to describe the subject project. If necessary, attach a detailed report.

Other Data:

Vehicle Occupancy (#)
___ AM ___ PM ___ 24-hour %
Percent by Transit:
___ AM % ___ PM % ___ 24-hour %
Percent by Carpool/Vanpool:
___ AM % ___ PM % ___ 24-hour %

Full-time Employees by Shift:

	Start Time	End Time	
First Shift:	___	___	Employees (#) ___
Second Shift:	___	___	Employees (#) ___
Third Shift:	___	___	Employees (#) ___

Parking Cost on Site: Hourly ___ Daily ___

Transportation Demand Management (TDM) Information:

At the time of this study, was there a TDM program (that may have impacted the trip generation characteristics of this site) under way?
- ☐ No
- ☐ Yes (If yes, please check appropriate box/boxes, describe the nature of this TDM program(s) and provide a source for any studies that may help quantify this impact. Attach additional sheets if necessary)

- ☐ (1) Transit Service
- ☐ (2) Carpool Programs
- ☐ (3) Vanpool Programs
- ☐ (4) Bicycle/Pedestrian Facilities and Site Improvements
- ☐ (5) Employer Support Measures
- ☐ (6) Preferential HOV Treatments
- ☐ (7) Transit and Ridesharing Incentives
- ☐ (8) Parking Supply and Pricing Management
- ☐ (9) Tolls and Congestion Pricing
- ☐ (10) Variable Work Hours/Compressed Work Weeks
- ☐ (11) Telecommuting
- ☐ (12) Other _____

Please Complete Form on Other Side

Institute of Transportation Engineers
Trip Generation Data Form (Part 2)

(All = All Vehicles Counted; Trucks = Heavy Duty Trucks & Buses)

Summary of Driveway Volumes

	Average Weekday (M-F)						Saturday						Sunday					
	Enter		Exit		Total		Enter		Exit		Total		Enter		Exit		Total	
	All	Trucks	All	Trucks	All	Trucks	All	Trucks	All	Trucks	All	Trucks	All	Trucks	All	Trucks	All	Trucks
24-Hour Volume																		
A.M. Peak Hour of Adjacent[1] Street Traffic (7 – 9) Time:																		
P.M. Peak Hour of Adjacent Street Traffic (4 – 6) Time:																		
A.M. Peak Hour: Generator[2] Time:																		
P.M. Peak Hour: Generator Time:																		
No. of Days Counted																		

1, 2. Please refer to the *Trip Generation User's Guide* for a definition of the terms.

Detailed Driveway Volumes—Average Weekday (M-F)

A.M. Period	Enter		Exit		Total		Mid-Day Period	Enter		Exit		Total		P.M. Period	Enter		Exit		Total	
	All	Trucks	All	Trucks	All	Trucks		All	Trucks	All	Trucks	All	Trucks		All	Trucks	All	Trucks	All	Trucks
6:00-6:15							11:00-11:15							3:00-3:15						
6:15-6:30							11:15-11:30							3:15-3:30						
6:30-6:45							11:30-11:45							3:30-3:45						
6:45-7:00							11:45-12:00							3:45-4:00						
7:00-7:15							12:00-12:15							4:00-4:15						
7:15-7:30							12:15-12:30							4:15-4:30						
7:30-7:45							12:30-12:45							4:30-4:45						
7:45-8:00							12:45-1:00							4:45-5:00						
8:00-8:15							1:00-1:15							5:00-5:15						
8:15-8:30							1:15-1:30							5:15-5:30						
8:30-8:45							1:30-1:45							5:30-5:45						
8:45-9:00							1:45-2:00							5:45-6:00						
9:00-9:15														6:00-6:15						
9:15-9:30														6:15-6:30						

Please attach any additional site information or comments regarding special site characteristics, if available.

☐ Check if additional information is attached.

Survey conducted by: Name: _____

Organization: _____

Address: _____

City/State/Zip: _____

Telephone #: _____ Fax #: _____ E-mail: _____

Please return to: Institute of Transportation Engineers
Programs & Services Department
525 School Street, SW, Suite 410
Washington, DC 20024-2797 USA
Telephone: +1 202/554-8050
FAX: +1 202/863-5486
ITE on the Web: http://www.ite.org

User Comments:

Trip Generation, 6th Edition

The Institute of Transportation Engineers would like to know what you think about the 6th Edition of *Trip Generation.* Please fill out the following questionnaire after you have had ample opportunity to use the new document. Your comments will help improve future editions of *Trip Generation.*

1. Please describe any errors or inconsistencies you have noted in this document. Please note page numbers and, if possible, a copy of the page(s) containing the error. Attach additional sheets if needed.

Description and page(s):_____

_____

_____

_____

_____

_____

_____

_____

_____

_____

_____

_____

_____

2. How easy to use and understand is the 6th Edition of *Trip Generation*?

    ❑ Very easy   ❑ Somewhat difficult
    ❑ Fairly easy   ❑ Very difficult

3. Please provide us with your comments, positive or negative, on the 6th Edition of *Trip Generation.*

_____

_____

_____

_____

_____

_____

_____

_____

4. Are there any specific enhancements or modifications that you would you like to see in future editions of *Trip Generation*?

_____

_____

_____

_____

5. For which additional land uses should the Institute collect trip generation data?

_____

_____

_____

_____

6. For specific land uses, which independent variables would you like to see added? Please specify the land use and the desired variable(s).

_____

_____

_____

_____

The following information is optional:

Name _____

Title _____

Agency or Firm _____

Address _____

City _____

State/Province _____ Postal Code _____

Country _____

Telephone: _____ Fax: _____ E-mail: _____

Thank you!

Please return this form to:
Institute of Transportation Engineers
Programs & Services Department
525 School St., S.W., Suite 410
Washington, D.C., 20024-2797 USA
Telephone: +1 202/554-8050
Fax: +1 202/863-5486
ITE on the Web: http://www.ite.org

User Comments:

Trip Generation, 6th Edition

The Institute of Transportation Engineers would like to know what you think about the 6th Edition of *Trip Generation.* Please fill out the following questionnaire after you have had ample opportunity to use the new document. Your comments will help improve future editions of *Trip Generation.*

1. Please describe any errors or inconsistencies you have noted in this document. Please note page numbers and, if possible, a copy of the page(s) containing the error. Attach additional sheets if needed.

Description and page(s):_____

_____

_____

_____

_____

_____

_____

_____

_____

_____

_____

_____

_____

2. How easy to use and understand is the 6th Edition of *Trip Generation*?

❑ Very easy   ❑ Somewhat difficult
❑ Fairly easy   ❑ Very difficult

3. Please provide us with your comments, positive or negative, on the 6th Edition of *Trip Generation.*

_____

_____

_____

_____

_____

_____

_____

_____

4. Are there any specific enhancements or modifications that you would you like to see in future editions of *Trip Generation*?

_____

_____

_____

_____

5. For which additional land uses should the Institute collect trip generation data?

_____

_____

_____

_____

6. For specific land uses, which independent variables would you like to see added? Please specify the land use and the desired variable(s).

_____

_____

_____

_____

The following information is optional:

Name _____

Title _____

Agency or Firm _____

Address _____

City _____

State/Province _____ Postal Code _____

Country _____

Telephone: _____ Fax: _____ E-mail: _____

Thank you!

Please return this form to:
Institute of Transportation Engineers
Programs & Services Department
525 School St., S.W., Suite 410
Washington, D.C., 20024-2797 USA
Telephone: +1 202/554-8050
Fax: +1 202/863-5486
ITE on the Web: http://www.ite.org

INDEX

Headings in plain type refer to land uses described in volumes 1 and 2. Subheadings (indented headings) refer to data tables and data plots for the independent variables associated with a specific land use.

Headings in **bold** type with page numbers preceded by the abbreviation "U" refer to topics described in volume 3, the *User's Guide*.

Bus Park-and-Ride Station
  See: Park-and-Ride Lot with Bus Service
Business Hotel 542-551
  Employees 543, 548-551
  Occupied Rooms 543-547
Business Park 1178-1193
  Acres 1189-1193
  Employees 1179-1183
  1000 Sq. Ft. Gross Floor Area 1184-1188

C

Campground/Recreational Vehicle
Park 630-635
  Acres 631
  Occupied Camp Sites 632-635
Car Wash
  *See: Gasoline/Service Station with Convenience
  Market and Car Wash*
Car Wash (Self-Service) 1483-1486
  Wash Stalls 1484-1486
Casino/Video Lottery Establishment 741-742
  1000 Sq. Ft. Gross Floor Area 742
Cemetery 941-948
  Acres 942-945
  Employees 942, 946-948
Church 900-909
  1000 Sq. Ft. Gross Floor Area 901-909
City Park 590-593
  Acres 591-592
  Picnic Sites 593
Clinic 1035-1042
  Employees 1036-1039
  Full-Time Doctors 1036, 1040-1041
  1000 Sq. Ft. Gross Floor Area 1036, 1042
Clothing Store
  See: Apparel Store
**Coefficient of Determination (R² Value) U-6,
U-14, U-18-19, U-22**
College (Junior/Community) 874-886
  Employees 875, 881-883
  1000 Sq. Ft. Gross Floor Area 875, 884-886
  Students 875-880
College/University 887-899
  Employees 894-899
  Students 888-893
Commercial Airport 4-31
  Average Flights per Day 14-22
  Commercial Flights per Day 23-31
  Employees 5-13
Community Center (Recreational) 792-801
  Employees 793
  Members 793
  1000 Sq. Ft. Gross Floor Area 793-801

Community/Junior College 874-886
  Employees 875, 881-883
  1000 Sq. Ft. Gross Floor Area 875, 884-886
  Students 875-880
Condominium/Townhouse (High-Rise
Residential) 393-402
  Dwelling Units 394-402
Condominium/Townhouse (Low-Rise
Residential) 388-392
  Dwelling Units 389-392
Condominium/Townhouse (Luxury) 403-407
  Occupied Dwelling Units 404-407
Condominium/Townhouse
(Residential) 360-387
  Dwelling Units 361-369
  Persons 370-378
  Vehicles 379-387
Congregate Care Facility 456-461
  Occupied Dwelling Units 457-461
Convenience Market
  *See: Gasoline/Service Station with
  Convenience Market*
Convenience Market (Open 15-16
Hours) 1540-1544
  1000 Sq. Ft. Gross Floor Area 1541-1544
Convenience Market (Open 24
Hours) 1530-1539
  1000 Sq. Ft. Gross Floor Area 1531-1539
Convenience Market with Gasoline
Pumps 1545-1562
  A.M. Peak Hour Traffic on Adjacent
    Street 1561
  1000 Sq. Ft. Gross Floor
    Area 1546, 1554-1560
  P.M. Peak Hour Traffic on Adjacent
    Street 1562
  Vehicle Fueling Positions 1546-1553
Corporate Headquarters Building 1059-1065
  Employees 1060-1062
  1000 Sq. Ft. Gross Floor Area 1063-1065
Correctional Facility
  See: Prison
County Park 594-603
  Acres 595-603

D

Day Care Center 913-940
  Employees 914-922
  1000 Sq. Ft. Gross Floor Area 923-931
  Students 932-940
Dependent Variable U-11, U-18-19
Detached Elderly Housing 452-455
  Dwelling Units 453-455

General Light Industrial 89-116
  Acres 108-116
  Employees 90-98
  1000 Sq. Ft. Gross Floor Area 99-107
General Office Building 1043-1058
  Employees 1045-1051
  1000 Sq. Ft. Gross Floor Area 1052-1058
GFA U-6, U-10-11
GLA U-6, U-10-11
Golf Course 675-702
  Acres 685-693
  Employees 676-684
  Holes 694-702
Golf Course (Miniature) 703-704
  Holes 704
Golf Driving Range 705-708
  Employees 706
  Tees/Driving Positions 707-708
Government Office Building 1092-1093
  Employees 1093
  1000 Sq. Ft. Gross Floor Area 1093
Government Office Complex 1132-1133
  Employees 1133
  1000 Sq. Ft. Gross Floor Area 1133
GRA U-6, U-11
Gross Floor Area U-6, U-10-11
Gross Leasable Area U-6, U-10-11
Gross Rentable Area U-6, U-11

H
Hardware/Paint Store 1252-1279
  Acres 1271-1279
  Employees 1253-1261
  1000 Sq. Ft. Gross Floor Area 1262-1270
Health Club 788-789
  1000 Sq. Ft. Gross Floor Area 789
Heavy Industrial (General) 117-131
  Acres 127-131
  Employees 119-123
  1000 Sq. Ft. Gross Floor Area 118, 124-126
High-Cube Warehouse 253-255
  Employees 255
  1000 Sq. Ft. Gross Floor Area 254
High-Rise Apartment 340-352
  Dwelling Units 342-350
  Persons 341, 351-352
High-Rise Residential Condominium/
Townhouse 393-402
  Dwelling Units 394-402
High School 846-873
  Employees 856-864
  1000 Sq. Ft. Gross Floor Area 865-873
  Students 847-855

High-Turnover (Sit-Down) Restaurant 1375-1393
  1000 Sq. Ft. Gross Floor Area 1376-1384
  Seats 1385-1393
Home Improvement Superstore 1591-1599
  1000 Sq. Ft. Gross Floor Area 1592-1599
Horse Racetrack 731-732
  Acres 732
  Employees 732
Hospital 977-1004
  Beds 996-1004
  Employees 978-986
  1000 Sq. Ft. Gross Floor Area 987-995
Hotel 502-529
  Employees 521-529
  Occupied Rooms 503-511
  Rooms 512-520
Hotel (All Suites) 530-541
  Employees 531
  Occupied rooms 532-536
  Rooms 537-541
Hotel (Business) 542-551
  Employees 543, 548-551
  Occupied Rooms 543-547
Hotel (Resort) 576-589
  Employees 577, 582-585
  Occupied Rooms 577-581
  Rooms 586-589

I
Ice Rink 739-740
  1000 Sq. Ft. Gross Floor Area 740
  Seats 740
Independent Variable U-6, U-9, U-11,
U-15, U-17-19, U-21-22
Industrial (General Heavy) 117-131
  Acres 127-131
  Employees 119-123
  1000 Sq. Ft. Gross Floor Area 118, 124-126
Industrial (General Light) 89-116
  Acres 108-116
  Employees 90-98
  1000 Sq. Ft. Gross Floor Area 99-107
Industrial Park 132-159
  Acres 151-159
  Employees 133-141
  1000 Sq. Ft. Gross Floor Area 142-150

J
Junior/Community College 874-886
  Employees 875, 881-883
  1000 Sq. Ft. Gross Floor Area 875, 884-886
  Students 875-880

Tire Store 1487-1503
  Employees 1499-1503
  1000 Sq. Ft. Gross Floor Area 1493-1498
  Service Bays 1488-1492
Tire Store (Wholesale) 1504-1518
  1000 Sq. Ft. Gross Floor Area 1512-1518
  Service Bays 1505-1511
Townhouse/Condominium (High-Rise
Residential) 393-402
  Dwelling Units 394-402
Townhouse/Condominium (Low-Rise
Residential) 388-392
  Dwelling Units 389-392
Townhouse/Condominium (Luxury) 403-407
  Occupied Dwelling Units 404-407
Townhouse/Condominium
(Residential) 360-387
  Dwelling Units 361-369
  Persons 370-378
  Vehicles 379-387
Townhouse (Rental) 358-359
  Dwelling Units 359
Toy/Children's Superstore 1606-1608
  1000 Sq. Ft. Gross Floor Area 1607-1608
Trip End U-9, U-11, U-18, U-21-22
Truck Terminal 55-74
  Acres 66
  Employees 57-65
  1000 Sq. Ft. Gross Floor Area 56
  Truck Berths 56

U

United States Post Office 1113-1131
  Employees 1114-1122
  1000 Sq. Ft. Gross Floor Area 1123-1131
University/College 887-899
  Employees 894-899
  Students 888-893
Utilities 256-261
  Acres 257, 261
  Employees 258-259
  1000 Sq. Ft. Gross Floor Area 257, 260

V

Vehicle Fueling Position U-11
VFP U-11
Video Arcade 1644-1645
  1000 Sq. Ft. Gross Floor Area 1645
  Video Games 1645

Video Lottery Establishment/Casino 741-742
  1000 Sq. Ft. Gross Floor Area 742
Video Rental Store 1646-1649
  Employees 1649
  1000 Sq. Ft. Gross Floor Area 1647-1648

W

Walk-in Bank 1650-1653
  Employees 1651
  1000 Sq. Ft. Gross Floor Area 1651-1653
Warehouse (High-Cube) 253-255
  Employees 255
  1000 Sq. Ft. Gross Floor Area 254
Warehouse (Mini) 216-252
  Acres 244-252
  Employees 217-225
  1000 Sq. Ft. Gross Floor Area 226-234
  Storage Units 235-243
Warehousing 188-215
  Acres 207-215
  Employees 189-197
  1000 Sq. Ft. Gross Floor Area 198-206
Water Slide Park 619-620
  Parking Spaces 620
Waterport/Marine Terminal 1-3
  Acres 3
  Berths 2
Wholesale Market 1570-1571
  Acres 1571
  Employees 1571
  1000 Sq. Ft. Gross Floor Area 1571
Wholesale Nursery 1308-1333
  Acres 1309, 1326-1333
  Employees 1309-1317
  1000 Sq. Ft. Gross Floor Area 1309, 1318-1325
Wholesale Tire Store 1504-1518
  1000 Sq. Ft. Gross Floor Area 1512-1518
  Service Bays 1505-1511

Z

Zoo 753-754
  Acres 754
  Employees 754